Low-cost
Sewerage

Low-cost Sewerage

Edited by

Duncan Mara
Department of Civil Engineering
University of Leeds, U.K.

JOHN WILEY & SONS
Chichester · New York · Brisbane · Toronto · Singapore

Copyright © 1996 by John Wiley & Sons Ltd.
Baffins Lane, Chichester,
West Sussex PO19 1UD, England

National   01243 779777
International (+44) 1243 779777

e-mail (for orders and customer service enquiries): cs-books@wiley.co.uk

Visit our Home Page on http://www.wiley.co.uk
or
http://www.wiley.com

Other Wiley Editorial Offices

John Wiley & Sons, Inc., 605 Third Avenue,
New York, NY 10158-0012, USA

Jacaranda Wiley Ltd, 33 Park Road, Milton,
Queensland 4064, Australia

John Wiley & Sons (Canada) Ltd, 22 Worcester Road,
Rexdale, Ontario M9W IL1, Canada

John Wiley & Sons (SEA) Pte Ltd, 37 Jalan Pemimpin #05-04,
Block B, Union Industrial Building, Singapore 2057

Library of Congress Cataloging-in-Publication Data

Low-cost sewerage / edited by Duncan Mara.
        p. cm.
    Includes bibliographical references and index.
    ISBN 0-471-96691-6 (alk. paper)
    1. Sewerage. I. Mara D. Duncan (David Duncan), 1944-
  TD653.L68   1996
  628.3—dc20
                                      96-15925
                                      CIP

British Library Cataloguing in Publication Data

A catalogue record for this book is available from the British Library

ISBN 0 471 966916

Typeset in 10½/13 Times by Acorn Bookwork, Salisbury

Contents

Preface

The International Conference on Low-Cost Sewerage was held at Weetwood Hall, University of Leeds, during 19–21 July 1995. Delegates attended from Colombia, Finland, Greece, India, Malaysia, Nigeria, Pakistan, South Africa, the United Kingdom, the United States, Venezuela and Zimbabwe. Fourteen papers were presented which, together with the Conference Conclusions, make up the 15 chapters of this book.

Low-cost sewerage is probably the only sanitation technology that is applicable in the high-density, low-income urban and periurban areas of developing countries. Often this will be simplified sewerage, especially its in-block variant now generally called condominial sewerage, although settled sewerage (which used to be called small-bore sewerage) may also be appropriate, particularly in areas already served by septic tanks. On-plot water supplies are commonly thought of as being a prerequisite for low-cost sewerage, but they are not essential: simplified sewerage worked well in a very poor slum area of Karachi which was served only by public standpipes.

Settled sewerage schemes were first installed in the late 1950s in what is now Zambia, and its inflective gradient design approach (which correctly ignores self-cleansing velocities) was developed in the United States in the mid-1970s. Simplified sewerage was a later development: condominial sewerage was first installed in low-income housing areas in Natal, northeast Brazil in the early 1980s, and it is now becoming quite common elsewhere in Brazil and the rest of Latin America.

Dissemination of the current state of knowledge about low-cost sewerage is extremely important if its literally vital message is to reach practising engineers in developing countries. The 1985 World Bank TAG Technical Note on small-bore sewerage and the 1986 UNCHS Manual on shallow (*i.e.* simplified) sewerage were a start. In 1994 the UNDP-World Bank Water and Sanitation Program produced design guidelines for simplified sewerage based on the minimum tractive tension (rather than self-cleansing velocity) design approach: this is now commonly used in southern Brazil and it needs to be more widely known.

I hope that this book will help to disseminate low-cost sewerage to a wider audience. If it succeeds in this, its publication will have been most worthwhile.

Duncan Mara
Leeds, February 1996

Contributors

Georgios E. Alexiou
AGP Consultants
Kleisouras 6
54631 Thessaloniki
Greece

Alexander Bakalian
UNDP—World Bank
Water and Sanitation Program
The World Bank
1818 H St NW
Washington, DC 20433
USA

Professor Georgios Balafutas
Department of Civil Engineering
Aristotle University of Thessaloniki
54006 Thessaloniki
Greece

David M. Brown
Department of Civil Engineering
Imperial College of Science,
   Technology and Medicine
London SW7 2BU

Dr David Butler
Department of Civil Engineering
Imperial College of Science,
   Technology and Medicine
London SW7 2BU

H.Brian Jackson
Engineering Division
Overseas Development Administration
94 Victoria Road
London SW1E 5JL

Peter J. Kolsky
Environmental Health Programme
London School of Hygiene & Tropical
   Medicine
Keppel Street
London WC1E 7HT

Professor D. Duncan Mara
Department of Civil Engineering
University of Leeds
Leeds LS2 9JT

Martin J. Marriot
Department of Civil Engineering
University of Hertfordshire
College Lane
Hatfield AL10 9AB

Richard N. Middleton
Kalbermatten Associates
2327 Pondside Terrace
Silver Spring, MD 20906
USA

Dr Richard J. Otis
Ayres Associates
2445 Darwin Road
Madison, WI 53705
USA

S. Ath. Panteliadis
AGP Consultants
Kleisouras 6
54631 Thessaloniki
Greece

Jon N. Parkinson
Environmental Health Programme
London School of Hygiene and
   Tropical Medicine
Keppel Street
London WC1E 7HT

Robert A. Reed
WEDC
Loughborough University of
   Technology
Loughborough LE11 3TU

José H. Rizo-Pombo
Carinsa
Apartado Aereo 118
Cartagena
Colombia

Professor John A. Swaffield
Department of Building Engineering
   and Surveying
Heriot-Watt University
Riccarton
Edinburgh EH14 4AS

Kevin Tayler
GHK International
St James Hall
Moore Park Road
London SW6 2JW

Dr Robin H.M. Wakelin
Department of Mechanical
   Engineering
Brunel University
Uxbridge UB8 3PH

Albert Wright
UNDP World Bank
Water and Sanitation Program
The World Bank
1818 H St NW
Washington, DC 20433
USA

1

Global Needs and Developments in Urban Sanitation

H. Brian Jackson

1.1 THE PROBLEMS

1.1.1 Rapid urbanization

By early in the twenty-first century, more than half of the world's population are predicted to be living in urban areas. By the year 2025 that proportion could rise to 60 per cent, comprising some 5 billion people. This rapid urban population growth is putting, and will continue to put, severe strains on the water supply and sanitation services in most major conurbations, especially those in developing countries. In certain major cities in Africa, for example, it is estimated that as many as two-thirds of the population are without adequate sanitation (Water Solidarity Network, 1994). This proportion is unlikely to reduce in the short-term as the peri-urbanization process, in which settlement precedes installation of basic services, is likely to be the dominant influence on urban growth for many years to come.

Inadequate water supplies alone will preclude the possibility of reliable, conventional, sewerage systems for many cities. Sewers can rapidly block if water is shut off for periods. It has been calculated that communities with waterborne sewerage normally require more than 75 litres per caput per day (lcd), compared with less than 20 lcd used in many squatter settlements (Cairncross and Feachem, 1993). Alternative sewerage technologies will increasingly be needed on grounds of water availability, construction skills and sustainability, as well as cost.

1.1.2 Inadequate appreciation of the full benefits of sanitary provision

Sanitation can be considered primarily as the safe removal of excreta, possibly with the inclusion of a partial treatment stage. There is a huge need for such sanitation, and associated hygiene practices, which can only begin to be met if they are accorded greater priority than hitherto.

Studies have indicated that, while improvement of water quality may lead to a 16 per cent reduction of infant diarrhoea, improved removal of excreta can lead to a 22% reduction (Esrey *et al.*, 1985). A 1990 review by USAID's Water and Sanitation for Health Programme of studies on the health effects of water and sanitation found that safe excreta disposal was the most effective intervention for reducing the incidence and severity of the six diseases studied (WASH, 1993).

Survey data from Guatemala indicate that children who lived in urban communities with poor sanitation were twice as likely to suffer from stunted growth due to bouts of diarrhoea as those who lived in communities with high levels of sanitation. In 'insanitary' areas of Bangladesh, the infant mortality rate is reported to be twice as high as that in the most affluent areas (WASH, 1993). These and similar findings are well documented, but their significance to planners and politicians is not yet universally accepted.

In western Europe during the nineteenth century the main effort to design and construct appropriate sanitation systems occurred in the wake of cholera, although in England the stench from the River Thames alongside the Houses of Parliament also had a significant influence upon politicians. In the words of one contemporary non-governmental organization (NGO) campaigner: 'Will we again have to wait until cholera epidemics in Latin America and elsewhere strike terror into cities and continents?' (Black, 1994). The author considers that 'in many rapidly urbanising countries, where life-threatening diarrhoeal disease is still endemic and erupts in periodic outbreaks, the urban sanitary crisis is a crisis simply waiting to happen'.

Notwithstanding this impassioned plea for rapidly expanded sanitation, the reality is that some communities are still likely to prefer water supply over sanitation and may be reluctant to pay for a facility where direct benefits are unclear. The demand for sanitation among target populations is therefore often low, and the interest of implementing agencies in investing in sanitation may also be low. This is due to a general lack of awareness of the health benefits from improved sanitation (LaFond, 1995).

1.1.3 Inadequate provision of sanitation facilities

The most detailed and authoritative data sets concerning the provision of sanitation in urban areas around the world are probably those given in the annual *Water Supply and Sanitation Sector Monitoring Reports* of the WHO/UNICEF Joint Water Supply and Sanitation Monitoring Programme (JMP). Commentators on the subject increasingly turn to the JMP reports for information on a country basis for the numbers of people served in total and by type of technology.

The UN Secretary General reported to the March 1994 session of the Committee on Natural Resources of the Economic and Social Council that, based on information received by the JMP from countries in the African and the Asian and Pacific regions, the percentage of the urban population with 'adequate' sanitation in 1993 was lower than had been estimated at the end of the International Drinking Water Supply and Sanitation Decade (IDWSSD) in 1990. More than 40 per cent of the urban population in Africa still did not have adequate sanitation, and in Asia and the Pacific the figure was 38 per cent (United Nations, 1994).

Drawing on earlier JMP data the World Bank's *World Development Report 1992* showed that a world total population in urban and rural areas of 1.7 billion were without adequate sanitation in 1990, and then projected that, even with accelerated investment and efficiency reforms, this number would not reduce before the year 2030 because of population growth (World Bank, 1992). With a 'business as usual' investment scenario the population without adequate facilities would increase to 3.2 billion by 2030.

The JMP report for 1993, giving the sector status as of 31 December 1991, still indicated a global sanitation-unserved population of 1.7 billion (WHO/UNICEF, 1993). This report also indicated that, in the 82 Third World countries monitored of 130 invited to participate, 60 per cent of the urban population (high income) that were served by sanitary facilities, had house connections to conventional sewerage. Septic tanks, pour-flush latrines and settled sewerage systems made up most of the balance, but no one group accounted for more than 15 per cent of the population. However, data from the preceding year, presumably for a somewhat different set of countries, indicated that conventional sewerage served only 40 per cent of the urban population, while settled sewerage and septic tanks were each catering for about 20 per cent of both the urban and the marginal urban populations (WHO/UNICEF, 1990; Water Solidarity Network, 1994). The trends in installation of low-cost sanitation systems clearly need further investigation.

The sanitation system terminology used by Cairncross and Feachem (1993) differs slightly from that used in the JMP, but a comparison of costs is made in the former reference which can also be related to the JMP categories. Based on a survey conducted by the World Bank in several developing countries the relative annual economic costs per household, according to Cairncross and Feachem (1993), were as follows:

Pour-flush toilets	10
Sewered pour-flush toilets	40
Conventional septic tanks	90
Conventional sewerage	100

The capital investment costs associated with the above alternative sanitation developments are generally expected to range from about US $75–150 to US $600–1200 per head of population in 1990 prices (Black, 1994).

1.2 INTERNATIONAL INITIATIVES TO IMPROVE SANITATION COVERAGE

1.2.1 Water Supply and Sanitation Collaborative Council

The IDWSSD did not meet its water supply and sanitation goals, largely due to the enormous increase in population during the 1980s. In recognition of this, and with a desire to ensure that Decade initiatives were not dissipated, various external support agencies met during the latter part of the 1980s to decide how the momentum could be maintained. At their 1990 meeting in New Delhi they decided that membership should be extended to include representatives from developing countries, and the resulting Water Supply and Sanitation Collaborative Council held its first meeting in Oslo in 1991.

Working groups were set up under the Collaborative Council, and a Sanitation Working Group produced its report *Sanitation: Unmet Challenge* for the Rabat meeting of the Council in 1993 with a strong plea for increased advocacy in the subject, stressing also the need for innovative cheaper technologies for safe excreta removal. Another working group was entrusted with the task of developing a strategy for the improved provision of water supply and sanitation services to urban areas. This group reported to the Barbados meeting of the Council in October 1995.

A continuing initiative of a further working group of the Council concerned with applied research, is GARNET (Global Applied Research Network—in water supply and sanitation). Although there are GARNET topic networks for many water supply aspects, and for sanitation topics such as nightsoil/sludge treatment, pit latrines and solid waste collection, so far there is none for sewerage systems.

1.2.2 UNDP–World Bank Water and Sanitation Program

Initiatives within the UN and World Bank system have included the UNDP/WB Water and Sanitation Program, which commenced with pilot projects in the late 1970s. This is a world-wide network dedicated to improving the access of poor people to safe water and sanitation on a sustainable basis. Working in over 40 developing countries, together with governments, donor agencies and NGOs, this Programme aims to promote 'innovative solutions tailored to meet local needs and conditions'. One of its more well known sanitation interventions has been in Kumasi, Ghana's second largest city, where, during the early 1990s, the programme helped significantly to improve both public and household latrines serving poor communities in part of the city, as well as to provide a basis for strategic sanitation planning for the whole city.

1.2.3 World Bank water resources policy

In its 1993 *Policy Paper on Water Resources Management* the World Bank recognized the economic and environmental consequences of inadequate sanitation, accepting that funding had received insufficient attention (World Bank, 1993). A review of 120 water supply and sanitation projects found that, while 104 projects funded water supply, only 58 included a sanitation component. In only a few of the cities with World Bank-financed water supply projects was adequate sewerage or sanitation provided to handle the increased wastewater created by the project. The World Bank policy paper reminds the reader that improved low-cost and more appropriate technologies are available to mitigate the high costs of conventional sewerage and sewage disposal systems.

The World Bank's *Water and Sanitation Sector Review* for the financial year 1993 (World Bank, 1994) argues that adoption of a demand-driven strategy could increase the level of lending for sanitation. Contrary to the findings of the WHO/UNICEF JMP reports, the World Bank claims that 'there is abundant evidence that urban families are willing to pay substantial amounts for the removal of excreta and wastewater from their neighbourhoods' (presumably on aesthetic grounds, even if the health benefits are insufficiently appreciated). It adds, however, that it is essential to create an enabling environment of conditions for the private sector, NGOs and consumers to play their parts in addition to that played by governments.

1.2.4 Rio Summit

The UN Conference on Environment and Development held in Rio de Janeiro in 1992 linked water and sustainable urban development within its concerns for the freshwater environment in chapter 18 of its agenda for the twenty-first century. It urged all states to introduce sanitary waste disposal facilities based on environmentally sound, low-cost and upgradable technologies, with a target of 75 per cent of the urban population to be provided with on-site or community sanitation facilities by the year 2000 (United Nations, 1992).

1.2.5 UNICEF

One of the subsequent developments to the Rio Summit has been a growth of activities by UNICEF in peri-urban areas compared with its traditional work among rural communities. During its Water and Environmental Sanitation Strategy Review Meeting in March 1995, UNICEF pledged to pay greater attention to sanitation in environmentally vulnerable urban areas.

1.3 INITIATIVES BY BILATERAL AID DONORS

Aid donors such as USAID, Overseas Development Administration (ODA), CIDA, DANIDA, SIDA, GTZ, DGIS of the Netherlands and SDC of Switzerland, and many others, have been well to the fore in both supporting UN and other international initiatives, and undertaking their own developmental assistance and research programmes for advancing the cause of urban sanitation. Two examples of donor programmes relevant to the main theme of this book are given below.

1.3.1 The USAID WASH Project

The WASH Project, which ran for 13 years before handing over to a wider-based environmental health project in 1993, is probably the premier activity within the sector. Among the many messages to be conveyed from the report *Lessons Learned in Water, Sanitation and Health* (WASH, 1993), I have selected the following:

> In the past, sanitary engineers simply assumed that a certain number of people would generate a certain volume of waste and it was the engineer's job to design and build a system large enough to handle that volume. Today, engineers are more inclined to examine where the waste is coming from, why it is being generated and how the volume and toxicity can be reduced. Whereas waterborne collection systems were once taken for granted as the norm, people now look for means of disposal that use less water.

> Tunisia, for example, with WASH assistance, developed and institutionalised a half dozen sanitation alternatives. Brazil has cut costs of urban sewers by 30 to 50 per cent by changing design norms of its sewer systems.

> 'Impossible situations' are more likely to be found in peri-urban communities, which are almost always built on cheap land with unattractive physical attributes, such as a steep slope or swampy conditions.

Low-cost technologies often require a high level of user maintenance—much higher than people in developed countries are accus-
tomed to. This implies a high level of community organisation and
participation.

1.3.2 UK ODA-funded Research

Recent funding for technology development and research by the
Engineering Division of the ODA has included studies of pit
emptying technologies, on-site sanitation in low-income urban
communities, urban surface-water drainage and reduced cost
sewerage systems, as well as wastewater treatment processes.

Experience from 10 countries, including the USA and
Australia, was considered in the reduced-cost sewerage study
which lasted over a period of 6 years in two phases, during
which time nine schemes were evaluated in Nigeria, Zambia,
Pakistan and Brazil. The study recommended that reduced-cost
sewerage should not be considered as a single sanitation concept,
but as a series of measures which, when applied to the principles
of conventional sewerage design and implementation, makes the
final product more appropriate to low-income communities in
developing countries (Reed, 1993*a*).

Three broad groups of low-cost systems were identified: simplified sewerage; condominial sewerage; and settled sewerage.
Measures were recommended for operation and maintenance,
and for overcoming institutional problems such as low connection rates and poor tariff collection. The report *Guidelines for
Reducing the Cost of Sewerage* was produced, which included
methods for capital cost reduction, sewer connection maximisa-
tion, system maintenance and system design optimization (Reed,
1993*b*). The results of many of the studies undertaken within the
above research project can be found in the recent publication
Sustainable Sewerage: Guidelines for Community Schemes (Reed,
1995).

The ODA-funded 1993 research report on reduced cost
sewerage concluded that it is unlikely that any sewerage scheme
can be reduced in cost to such an extent that it can be fully paid
for by the urban poor, although it should be feasible to recover
the costs of operation and maintenance (Reed, 1993*a*). Even the
interim report of 1989 was able to quote 84 references from

similar studies and reports, following on from the publication of the World Bank TAG Technical Note *The Design of Small Bore Sewer Systems* (Otis and Mara, 1985) and the UNCHS report *The Design of Shallow Sewer Systems* (Sinnatamby, 1986). With such an interest it is pertinent to ask why low-cost sewerage is not in even greater use, especially in developing countries.

1.4 INITIATIVES AT COUNTRY LEVEL

Probably the most well known, and well documented, local initiative in meeting the demand for sanitation services was with the Orangi Pilot Project (OPP) in Karachi, Pakistan (see Serageldin, 1994). Local people aspired to a traditional full sewerage system as they already had a relatively satisfactory water supply. As this was financially impracticable, they were then persuaded within the OPP to contribute labour in return for financial assistance towards the construction of in-house sanitary latrines, house connections and lane sewers. As the power of the OPP-related organizations increased, they were able to bring pressure on the municipality to provide funds for the contribution of secondary and primary sewers, ultimately benefiting more than 600 000 poor people. This experience demonstrated how people's demands move naturally from the provision of water to removal of waste from their homes, then from their blocks, and finally from their neighbourhood.

Other local initiatives can be found throughout Asia, with the NGO sector in Bangladesh being particularly active in working with local authorities to provide better facilities. The UK NGO, WaterAid, has a similar story to tell in Bangalore India, albeit on a smaller scale than Orangi. The ODA assistance with slum improvement projects in Indian cities such as Indore is building on local systems and preferences for waste removal, as well as drawing on the experience of its research-funded urban drainage activities.

Several countries in South America have interesting records of local initiatives, particularly concerning low-cost condominial sewerage, with examples from Brazil and Argentina being perhaps the most well known. Within the city of São Paulo, for example, a small local municipal agency was able to demonstrate to the state water utility, through its experimentation with new technical and institutional ways of providing water and sanitation, that the provision of such services to the low-income, informal settlements (*favelas*) was possible: in 1980 less than 1 per cent of the *favela* population had a sewerage system, whereas by 1990 this had increased to 15 per cent (Serageldin, 1994). In the Buenos Aires barrio of Martin Coronado a residential cooperative was responsible for many technological sanitation innovations as well as a sewer network of more than 60 km of domiciliary collectors (MAE/DGCS and CEFRE, 1993).

In African countries, including Kenya, Nigeria, Ghana and Zambia, there has been much interest in sewerage for aqua privies which are a common form of latrine in many areas. In some of the new Egyptian townships near the Suez Canal local designs for septic tanks and small-bore settled sewers are being used in conjunction with prototype gravel bed hydroponic wastewater treatment works based on donor-funded research in Egypt.

In Kumasi in Ghana on-site sanitation systems provide the services to most of the inhabitants in both the indigenous low-cost living areas and high-cost new government housing areas. As the housing density is not high, there is space for drainage fields, which are hydrogeologically satisfactory. However, for tenement areas, which house about 25 per cent of the population, the only technically feasible solution for the proper disposal of excreta and sullage was found to be sewerage (Obeng and Locussol, 1992). Capital cost comparisons for collection and treatment of the three sewerage options of simplified sewerage, settled sewerage and conventional sewerage, led to the conclusion that the most affordable and least-cost solution for the Kumasi tenement areas was simplified sewerage. Surprisingly, however, the difference in cost between this and conventional sewerage was not large (US$15m versus US$18m). The nature of the terrain in this instance reduced the potential cost saving of the former over the latter.

Further interesting findings from Kumasi, based on willingness-to-pay surveys undertaken by local officials, were that (MAE/DGCS and CEFRE, 1993):

- families on average were willing to pay about the same amount for sanitation as they paid for rent, electricity or water;

- the poorest people who used public latrines were spending more for sanitation than those with household systems;

- people were willing to pay approximately the same for a household latrine as for a sewer connection; and

- the poor were willing to pay for latrines, but not for sewerage because of low reliability in the past.

1.5 CONCLUSIONS

Various low-cost sewerage initiatives are at work throughout the world, especially in countries with water shortages and with large urban poor populations. However, the advantages of so-called lower-cost sewerage systems over the so-called conventional systems are not well known to the wider public, as the evidence of demand for such services and confidence in their reliability is conflicting. The political and financial will to give urban sanitation a higher profile is therefore patchy.

It is my hope that the proceedings of this Conference will provide a valuable exchange of ideas and experience to enhance the opportunity for greater provision of appropriate, safe and sustainable sanitation to as wide an urban population as possible.

1.6 REFERENCES

Black, M. (1994). *Mega-Slums: The Coming Sanitary Crisis.* London: WaterAid.

Cairncross, S. and Feachem, R. (1993). *Environmental Health Engineering in the Tropics,* 2nd edn. Chichester: John Wiley.

Esrey, S.A., Feachem, R.G. and Hughes, J.M. (1985). Interventions of the control of diarrhoeal diseases among young children: improving water supplies and excreta disposal facilities. *Bulletin of the World Health Organization,* **63**, 757–772.

LaFond, A. (1995). *A Review of Sanitation Program Evaluations in Developing Countries.* EHP Activity Report No. 5. Arlington, VA: Environmental Health Project.

MAE/DGCS and CEFRE (1993). *Working Group on Urbanization Report to WSSCC Rabat Meeting (7–10 September).* Rome: Ministry of Foreign Affairs.

Obeng, L.A. and Locussol, A.R. (1992). *Sanitation Planning: A Challenge for the 90's.* Abidjan: The World Bank (West Africa Water and Sanitation Program).

Otis, R.J. and Mara, D.D. (1985). *The Design of Small-bore Sewer Systems.* TAG Technical Note No. 14. Washington, DC: The World Bank.

Reed, R.A. (1993*a*). *Reduced Cost Sewerage for Developing Countries*: *Phase II (Final Report)*. Loughborough: University of Technology (WEDC).

Reed, R.A. (1993*b*). *Guidelines for Reducing the Cost of Sewerage*. Loughborough: University of Technology (WEDC).

Reed, R.A. (1995). *Sustainable Sewerage: Guidelines for Community Schemes*. London: IT Publications.

Serageldin, I. (1994). *Water Supply Sanitation and Environmental Sustainability: The Financing Challenge*. Washington, DC: The World Bank.

Sinnatamby, G. S. (1986). *The Design of Shallow Sewer Systems*. Nairobi: UNCHS.

United Nations (1992). *Report of the UN Conference on Environment and Development (Rio de Janeiro, 3–14 June)*, Vol. 1, Chapter 18. New York: United Nations.

United Nations (1994). *Water Resources: Progress in the Implementation of the Mar del Plata Action Plan and of Agenda 21 on Water-related Issues*. Report of the Secretary General to the Economic and Social Council, 12 January. New York: United Nations.

WASH (1993). *Lessons Learned in Water, Sanitation and Health: Thirteen Years of Experience in Developing Countries*. Arlington, VA: Water and Sanitation for Health Project.

Water Solidarity Network (1994). *Water and Health in Underprivileged Urban Areas*. Paris: Water Solidarity Network.

WHO/UNICEF (1990). *Water Supply and Sanitation Sector Monitoring Report 1990 (Baseline Year)*. Geneva: World Health Organization.

WHO/UNICEF (1993). *Water Supply and Sanitation Sector Monitoring Report 1993 (Sector Status as of 31 December 1991)*. Geneva: World Health Organization.

World Bank (1992). *World Bank Development Report 1992: Development and the Environment*. New York: Oxford University Press.

World Bank (1993). *Water Resources Management*. World Bank Policy Paper. Washington, DC: The World Bank.

World Bank (1994). *Annual Review of Portfolio Performance FY93— Water and Sanitation Sector*. Report No. TWU-OR2. Washington, DC: The World Bank.

2

Unconventional Sewerage Systems: Their Role in Low-cost Urban Sanitation

D.D. MARA

2.1 INTRODUCTION

The health of low-income communities in peri-urban areas of developing countries is extremely precarious. Infant and under-5 mortality rates are higher in peri-urban areas than rural areas, and these high rates of mortality, and also morbidity, are due to poverty, but they more directly reflect the gross inadequacies of peri-urban infrastructure, especially water and sanitation services.

Urban health in *all* countries is closely linked to high levels of water supply, sanitation and personal and domestic hygiene. Similarly, peri-urban ill-health in developing countries is a direct result of inadequate water supplies, inadequate sanitation facilities, inadequate collection of solid wastes and consequently low levels of personal and domestic hygiene. The World Health Organization (WHO) estimates that as much as 80 per cent of all morbidity in developing countries is due to water- and excreta-related diseases, i.e. a result inadequate water supplies and inadequate sanitation. If the peri-urban environment in developing countries is not changed, people will continue to be ill and, in the words of the late Barbara Ward, will 'continue to defecate themselves to death.'

The International Drinking Water Supply Decade (1981–90) strove, and its successor Safe Water 2000 (1991–2000) is striving, towards the goal of adequate water and adequate sanitation for all. However, population growth, coupled with insufficient investment, makes this an uphill struggle. None the less governments, together with water supply and sanitation engineers, must continue to develop programmes that improve urban, especially peri-urban, health. Low-cost sewerage has a vital role in this area of global public health.

2.2 ON-SITE OR OFF-SITE SANITATION?

The research and development done by the World Bank during 1976–86 (see Kalbermatten *et al.,* 1982*a,b*; Feachem *et al.,* 1983) has clearly shown that possession and proper use and maintenance of a sanitation facility is more important, in terms of improving health, than the actual sanitation technology employed, provided of course that it is affordable and socio-culturally acceptable. None the less, sanitation technology choices have to be made, and the principal choice is between on-site and off-site systems. These are the following:

On-site technologies: VIP latrines
                              Pour-flush toilets
                              Septic tanks

Off-site technologies: Conventional sewerage
                              Unconventional sewerage
                              • settled sewerage
                              • simplified sewerage

These are adequately described in the literature—see, for example, Mara (1984, 1985), Mara and Sinnatamby (1986), Otis and Mara (1985), Sinnatamby (1986), Bakalian *et al.* (1994) and, more generally, Mara (1996). On-site systems can be upgraded over time, and with corresponding improvements in water supply, to settled sewerage systems (see Kalbermatten *et al.,* 1982*b*; Mara, 1996).

2.2.1 Nomenclature of unconventional sewerage

At present this is confusing: both the unconventional sewerage technologies listed above use small diameter sewers laid at shallow depths and in which the flow is, wherever possible, by gravity; so terms such as small-bore sewerage, small diameter gravity sewerage or shallow sewerage are unclear. Following the Portuguese terminology developed in Brazil (see Guimarães, 1986), I wish to propose the following:

1. *Settled sewerage,* to describe the system in which wastewater from one or more households is discharged into a single-compartment septic tank (usually termed in this context a solids interceptor tank), the settled (or solids-free) effluent from which is discharged into shallow, small-bore gravity sewers. Settled sewerage is thus small-bore sewerage *sensu* Otis and Mara (1985), small diameter gravity sewerage *sensu* Otis (Chapter 9) and common effluent drainage *sensu* South Australian Health Commission (1982). In Portuguese it is called *redes de esgotos decantados*; in French, *réseaux d'eaux usées decantées*; and in Spanish (Rizo Pombo, Chapter 10), *acantarillado sin arrastre de sólidos.*

2. *Simplified sewerage,* to describe shallow sewerage *sensu* Sinna-tamby (1986) and its in-block variant sometimes called backyard or condominial sewerage (Rodrigues de Melo, 1985). This system does not convey presettled sewage, but is essentially conventional sewerage without any of its conservative design requirements that have accrued over the past century or so; it can be considered as the latter stripped down to its hydraulic basics. In Portuguese, it is called *redes de esgotos simplificadas*; in French, *réseaux d'eaux usées simplifiés*; and in Spanish, *accantarillado simplificado.*

2.2.2 The case for on-site systems

The development of simplified sewerage in northeast Brazil in the early 1980s (see Sinnatamby, 1983) has really changed our views on the role of on-site systems. Despite their undoubted technical feasibility (see Roy *et al.*, 1984; Middleton, 1995), they can now only be considered appropriate in urban/peri-urban areas if they are less expensive than simplified sewerage. The reason for this is that above a certain population density (160 persons per ha in the case of Natal in northeast Brazil on which Figure 2.1 is based) simplified sewerage is cheaper than on-site systems. Thus sanitation engineers should now always ascertain in every case

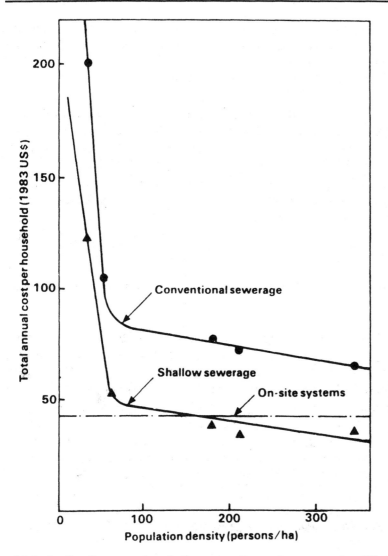

Figure 2.1 Costs of conventional sewerage, shallow (i.e. simplified and, in fact, condominial) sewerage and on-site sanitation as a function of population density in the City of Natal, Rio Grande do Norte, northeast Brazil (from Sinnatamby, 1983, 1986)

which is the lower-cost solution: on-site systems or simplified sewerage. Lower cost in this context means both economic and financial, and must include operation and maintenance costs as well as investment costs.

2.2.3 The case for settled sewerage

Here also the case has to be made on economic or financial grounds: is settled sewerage cheaper than on-site systems or simplified sewerage? If the community already has septic tanks, then (assuming the soil can no longer accept the septic tank effluent—but see below) probably settled sewerage will be cheaper than simplified sewerage; but, of course, this needs to be checked in each case.

However, if the reason why the soil can no longer accept septic tank effluent is simply because of a high, in-house water consumption (>100 litres/caput day), with a correspondingly high wastewater generation, then serious consideration should be given to in-house water conservation techniques, such as the installation of water-saving plumbing fixtures (see Mara, 1989), in order to reduce the resulting wastewater flow such that the soil is again able to accept the septic tank effluent.

2.2.4 The case for simplified sewerage

As will be now apparent, there is a general case in low-cost urban sanitation programmes and projects, especially those in high-density settlements, for considering simplified sewerage as the sanitation technology of first choice. It needs to be confirmed, therefore, in each situation whether simplified sewerage is:

● cheaper than on-site sanitation, and

● cheaper than settled sewerage

Simplified sewerage would, therefore, only *not* be finally selected if it were in fact *not* cheaper than either of these alternatives. This is only likely at low-population densities or in areas already served by septic tanks (even currently malfunctioning septic tanks).

A final decision to make concerning simplified sewerage is whether to adopt condominial (or backyard) sewerage or in-street sewerage. The former is more generally favoured in northeast Brazil, for example, and the latter in southern Brazil where SANEPAR, the water and sewerage company of the State of Paraná, often installs 'double sewers', i.e. a sewer on each side of the street under each sidewalk. It is not clear whether the reasons for this are always really valid, but what is clear is that 'double in-street simplified sewerage' is significantly (about two-thirds) more expensive than condominial sewerage (Hamer, 1995; Mara, 1996).

2.3 REFERENCES

Bakalian, A., Wright, A., Otis, R. and de Azevedo Netto, J. (1994). *Simplified Sewerage: Design Guidelines.* Water and Sanitation Report No.7. Washington, DC: The World Bank.

Feachem, R.G., Bradley, D.J., Garelick, H. and Mara, D.D. (1983). *Sanitation and Disease: Health Aspects of Excreta and Wastewater Management.* Chichester: John Wiley.

Guimarães, A.S.P. (1986). *Redes de Esgotos Simplificadas.* Brasília: Programa das Nações Unidas para o Desenvolvimento/Ministério do Desenvolvimento Urbano e Meio Ambiente.

Hamer, J. (1995). *Low Cost Urban Sanitation in Developing Countries.* B.Eng dissertation. Leeds: University of Leeds (Department of Civil Engineering).

Kalbermatten, J.M., Julius, D.S. and Gunnerson, C.G. (1982a). *Appropriate Sanitation Alternatives: A Technical and Economic Appraisal.* Baltimore: Johns Hopkins University Press.

Kalbermatten, J.M., Julius, D.S., Gunnerson, C.G. and Mara, D.D. (1982b). *Appropriate Sanitation Alternatives: A Planning and Design Manual.* Baltimore: Johns Hopkins University Press.

Mara, D.D. (1984). *The Design of Ventilated Improved Pit Latrines.* TAG Technical Note No. 13. Washington, DC: The World Bank.

Mara, D.D. (1985). *The Design of Pour-Flush Latrines.* TAG Technical Note No. 15. Washington, DC: The World Bank.

Mara, D.D. (1989). *The Conservation of Drinking Water Supplies: Techniques for Low-income Settlements.* Nairobi: United Nations Centre for Human Settlements.

Mara, D.D. (1996). *Low-cost Urban Sanitation.* Chichester: John Wiley.

Mara, D.D. and Sinnatamby, G.S. (1986). Rational design of septic tanks in warm climates. *The Public Health Engineer,* **14**(4), 49–55.

Middleton, R.N. (1995). Making VIP latrines succeed. *Waterlines,* **13**(4), 27–29.

Otis, R.J. and Mara, D.D. (1985). *The Design of Small Bore Sewer Systems.* TAG Technical Note No. 14. Washington, DC: The World Bank.

Rodrigues de Melo, J.C. (1985). Sistemas condominiais de esgotos. *Engenharia Sanitária (Rio de Janeiro),* **24**(2), 237–238.

Roy, A.K. et al. (1984). *Manual on the Design, Construction and Maintenance of Low-cost Pour-flush Waterseal Latrines in India.* TAG Technical Note No. 10. Washington, DC: The World Bank.

Sinnatamby, G.S. (1983). *Low-cost Sanitation Systems for Urban Peripheral Areas in Northeast Brazil.* PhD thesis. Leeds: University of Leeds.

Sinnatamby, G.S. (1986). *The Design of Shallow Sewer Systems.* Nairobi: United Nations Centre for Human Settlements.

South Australian Health Commission (1982). *Common Effluent Drainage System.* Adelaide: SAHC (Health Surveying Services).

3

Selecting Communities for Sewerage

Robert A. Reed

3.1 INTRODUCTION

In most countries the demand for sewerage greatly exceeds the resources available to provide it. Factors such as finance, engineering skills and administrative capability all limit the number of new sewerage schemes that can be initiated at one time. Even though resources are so limited it is a sad fact that many of the communities that have received sewerage underutilize it. This causes operational problems and poor return on investment. Low returns from the community lead to cut backs in maintenance and reductions in the funds available for future schemes. There is a need therefore to select the schemes which should receive priority and should be implemented first.

In the past many schemes have been selected on an *ad hoc* basis with more attention being paid to the political and economic strength of the recipient communities than whether the schemes are likely to be sustainable, cost-effective or needed. International donors and financial institutions are aware that the failure of many infrastructure projects is because of poor project identification, and this has led to a demand for a more rigorous approach to scheme selection.

This chapter suggests a mechanism for objectively selecting which, of a group of communities, should be the first to receive sewerage. It is divided into two sections. The first section discusses the criteria affecting the need for and prioritization of sewerage schemes. Topics including population density, industrial pollution, affordability and groundwater pollution are discussed. Failure of on-site sanitation systems is given particular emphasis as this is the most common reason given for requiring sewerage. The objective measurement of each of the criterion is stressed.

The second section suggests a mechanism for objectively prioritizing communities for sewerage, using the selection criteria discussed in section one. A weighted matrix approach is suggested. Each community is awarded a score for each of the criteria and that score is multiplied by the weight assigned to that topic. Criteria considered to have most influence on the prioritizing process receive the highest weight. The weighted scores of each community are then totalled and the communities having the highest total score are most favoured for sewerage. The paper uses a worked example to explain the approach. The results of the matrix are finally tested with a sensitivity analysis to asses the confidence of the predictions.

3.2 CRITERIA AFFECTING THE SELECTION OF SEWERAGE SCHEMES

The list of factors affecting the prioritization of sewerage schemes is long and will vary from place to place and over time, depending on the conditions prevailing when the analysis is being done. The criteria described below are the most common, but planners should feel free to add or subtract criteria to suit local conditions. As an example, the author undertook a selection analysis in which pollution of coastal lagoons was considered a critical factor. Such a consideration would only be necessary for communities adjacent to lagoons and reefs of national importance. In the majority of situations such a criterion would have little relevance.

As with most design activities, current conditions are only of importance as a guide to what is likely to happen in the future. Sewerage schemes are designed to meet the needs of a community for many years (called the design life) and must be capable of handling the maximum sewage flow, which usually comes at the end of the design life. In general, selection criteria should relate to the conditions expected at the end of the design life rather than those currently experienced. If, however, there are current conditions that are relevant to the prioritization process (such as heavy surface water pollution or schemes proposed for built-up areas, which have already reached optimum density) then they should be included.

It is important to remember that all selection criteria must be measurable. The whole purpose of the exercise is to make selection as objective as possible and minimize subjective decision-making. Criteria such as 'consumer demand' should only be used if based on a consumer survey carried out and analysed by a competent organization. The following criteria could form a basis for project selection.

3.2.1 Projected total population

Schemes expected to serve large populations at the end of the design life are generally considered to have a higher priority than those with a lower projected population. This is because large schemes tend to have a lower per capita cost, produce greater social and environmental gains, maximize the number of people having access to improved sanitation and improve the local environment through reductions in odour and inconvenience from open sullage drains.

3.2.2 Population density

Generally, the higher the population density, the greater the health hazard from poor sanitation and the lower the unit cost of sewerage. Also, as housing density increases and plot sizes decrease the chances of on-site sanitation systems failing increases.

3.2.3 Failure of on-site sanitation systems

The most common types of on-site system are pit latrines (of varying types) and septic tanks. Readers are advised to consult Franceys *et al.* (1992) for details of the design and construction of such systems.

The failure of existing on-site sanitation systems is one of the commonest reasons given for needing sewerage. As sewerage is far more expensive than on-site sanitation the planner should check the causes of any failure (current or projected) before altering the *status quo*. In many cases failure occurs because of poor design, construction or operation and in such cases, renovation of on-site systems will nearly always be more appropriate than their replacement by sewerage. During a visit to Mauritius the author was asked to comment on a proposal to provide sewerage to a number of communities where on-site sanitation was said to have failed (STWI, 1993). Site investigation showed that the reason the pit latrines were no longer working was that the pits were full. A programme of pit emptying or replacement was a much cheaper alternative to sewerage. Sewerage should only be considered in communities where on-site sanitation can be proved to have failed irreparably or is likely to fail within the design life (Reed, 1994).

3.2.4 Industrial pollution

Although it is rare for a public sewerage scheme to be constructed purely for the treatment of industrial pollution, it is often a significant contributory factor in deciding which community should be sewered. Industrial effluent can be much more concentrated than domestic sewage and it can produce serious pollution of water sources. Some industries produce large volumes of effluent and its disposal to a communal sewerage network may require sewer pipe diameters to be increased.

If some of the communities being considered for sewerage are industrialized, it will be necessary to find a method for including the impact of those industries on the selection process. A comparison of the organic load (in terms of kilograms of BOD per day) produced by industry is one method, but this is more important where sewage treatment is to be included in the scheme. Effluent volume and discharge times may well be more relevant because of their effect on sewer pipe sizes.

3.2.5 Cost

Some measure of the cost of a scheme must be included in the selection procedure. A comparison of total capital cost (assuming all the schemes are designed using the same criteria) will identify which schemes are the cheapest and which are within the budget available. It will not show how effectively the money is being spent. A better measure is the unit capital cost of a scheme. Communities where the wastes are primarily domestic can be measured in terms of the cost of the scheme per person served. Where there is a significant industrial or commercial load the cost per cubic metre of effluent may be more useful.

Operational costs must also be considered. Most sewerage networks operate under gravity and therefore the operational costs are approximately proportional to the size of the network. Schemes including pumping will have higher operational costs and this could be taken into account in the selection process.

Most of the factors so far discussed are technical or financial. There will often be other factors relevant to a selection process and peculiar to the particular circumstances. When considering them, it is important that some form of objective measuring and/ or estimation criteria are developed.

3.2.6 Tourist impact

In many countries tourism is a major source of employment and foreign exchange. The impact on tourists of unsightly polluted drains, bad odours and 'non-standard' sanitary fixtures can far exceed the potential health hazard. This makes many community leaders consider the provision of sewerage a priority in areas where tourism is or could be an important activity. Comparison between communities having a tourist interest is difficult; numbers of tourists, length of polluted drains, length of visible polluted drains or pollution of beaches and bathing waters are all possible criteria.

3.2.7 Environmental impact

The impact of a sewerage scheme on the environment may be both positive and negative. Reductions in environmental odour and unsightly pollution within the community may be counterbalanced by increased pollution concentration in the receiving waters with consequent reductions in flora and fauna. Objective evaluation of environmental impact is expensive and requires detailed surveys of the current situation. In this case it is usually sufficient to rely on informed professional judgement.

3.2.8 Affordability

In countries where the demand for sewerage far outstrips the government's ability to provide it, selection based purely on need is not enough. Any long-term policy to provide a service must be based on certain assumptions about how much income the provider will receive from its customers. Where there is equal need, it may only be possible in the first instance to serve communities with a known ability and willingness to pay necessary tariffs, tackling those where higher subsidies may be needed at a later date.

A community's ability to pay for a service can be assessed by comparing the likely tariff with the minimum income levels of the majority of the community. It is normally accepted that a family should pay no more than 2 per cent of its income on sanitation. If a measure of cross-subsidy is assumed within the community, tariffs should be less than 2 per cent of the income of the top 80 per cent of families.

A community's ability to pay for sewerage is not the same as a willingness to pay for the service. Willingness to pay is related more to the perceived importance of the service than to its cost. In communities where sewerage is a high priority, there may be a willingness to contribute a higher percentage of income than the 2 per cent mentioned previously. Conversely, communities having little desire for sewerage will be unwilling to pay 2 per cent of their income for a service for which they have little demand. Estimates of how much a community is willing to pay for sewerage can be obtained from an analysis of household surveys but this can only be a snapshot of current opinion. A community's opinion of what is a reasonable tariff will change over time, depending on changes in general economic conditions, the length of time a service has been in operation, and the services offered to surrounding communities. It is probably safer to make decisions on cost-recovery based on ability to pay rather than willingness to pay.

3.2.9 Economy of scale

If a number of the communities are close to each other or close to an existing scheme, there may be some economy of scale if a single treatment plant or outfall sewer can be constructed to serve them all.

3.2.10 Institutional capacity

Sewerage schemes are predominantly implemented, operated and maintained by institutions. Often an institution exists before a scheme commences and it is expected to expand to encompass the new duties necessary for a scheme's construction and operation. An institution's ability to cope with the demands of a new sewerage scheme will greatly affect that scheme's long-term success, both technically and financially. Communities having institutions with the capacity to encompass new commitments and responsibilities efficiently and competently should be favoured in any selection process over communities having institutions that cannot. Decisions on an institution's capacity to cope with added commitments should be based on an institutional evaluation.

3.2.11 Health benefits

One of the main reasons for considering sewerage is to improve standards of health, although not all schemes will produce the same health benefits. Improvements in health will depend on the existing levels of health which, in turn, will partially depend on current levels of sanitation and hygiene practice. It is rarely possible to quantify improvements in health resulting from the provision of sewerage because of the number of variables involved. In most situations it may be assumed that similar schemes will produce similar health benefits.

3.3 NUMERICAL ANALYSIS OF NEED AND VIABILITY CRITERIA

Having decided which criteria affect the ranking of a group of proposed sewerage schemes it is necessary to determine which of the schemes should have priority for implementation. This chapter recommends a numerical analysis as follows:

- each community is awarded a 'score' for each of the criteria included in the ranking process. For each criterion, a high score indicates that the provision of sewerage to that community is important while a low score indicates that it is less critical;

- not all the criteria included are of equal importance. It is necessary to weight the scores so that important criteria have a larger impact on the final result than minor ones;

- a weighted score is produced by multiplying the score by the weight;

- the total score for a community is the sum of the weighted scores given for each criterion; and

- communities having the highest scores are most favoured for the immediate implementation of a sewerage scheme.

The magnitude of the scores is unimportant: only the ranking of the communities matters. An example of the selection process is shown in Table 3.1. The table concentrates on technical and

Table 3.1 Example of a matrix for prioritising the provision of sewerage

Town	Projected population			On-site sanitation failure			Polluting industries			Unit cost of construction		
	Sc	Wt	T	Sc	Wt	T	Sc	Wt	T	Sc	Wt	T
A	10	1	10	1	3	3	1	1	1	4	2	8
B	9	1	9	7	3	21	10	1	10	7	2	14
C	6	1	6	1	3	3	10	1	10	5	2	10
D	3	1	3	1	3	3	1	1	1	2	2	4
E	2	1	2	1	3	3	1	1	1	3	2	6
F	2	1	2	1	3	3	10	1	10	5	2	10
G	2	1	2	1	3	3	5	1	5	2	2	4
H	2	1	2	1	3	3	5	1	5	1	2	2
J	1	1	1	4	3	12	10	1	10	10	2	20
K	1	1	1	1	3	3	1	1	1	2	2	4
L	1	1	1	7	3	21	10	1	10	6	2	12

Sc = score (the impact of a criterion on a town).
Wt = weight (the importance of a criterion in the selection process).
T = total (the multiple of the score and the weight).

financial criteria. It is assumed that other considerations such as health and the environment, which could be included, will be dealt with separately. It is also assumed that the schemes are all designed and constructed using the same criteria.

3.3.1 Scoring for individual criteria

The range of the scoring system for individual criteria and whether the scale from high priority to low priority is ascending or descending is immaterial provided a consistent system is used for all criteria. In the example shown in Table 3.1, a scale of 1–10 has been used, with 1 being minimum importance and 10 being maximum. Other ranges are equally valid but they should reflect the level of accuracy with which criteria can be measured, bearing in mind that all the criteria must be measured using the same scale. As an example, the method used for determining the scores for each of the criteria used in Table 3.1 now follows.

The highest and lowest numbers in the range shown in Table 3.2 correspond approximately to the largest and smallest commu-

Tourist impact			Consumers' ability to pay			Annual unit O & M cost			Total score
Sc	Wt	T	Sc	Wt	T	Sc	Wt	T	
1	3	3	4	2	8	5	1	5	38 (8)
4	3	12	7	2	14	10	1	10	90 (1)
4	3	12	7	2	14	10	1	10	65 (5)
10	3	30	10	2	20	5	1	5	66 (4)
2	3	6	4	2	8	10	1	10	36 (10)
3	3	9	7	2	14	10	1	10	58 (6)
4	3	12	4	2	8	5	1	5	39 (7)
4	3	12	4	2	8	5	1	5	37 (9)
5	3	15	10	2	20	1	1	1	79 (2)
1	3	3	1	2	2	10	1	10	24 (11)
3	3	9	7	2	14	10	1	10	77 (3)

Table 3.2 Projected population

Population range	Score	Population range	Score
15 000–16 300	1	21 500–22 800	6
16 300–17.600	2	22 800–24 100	7
17 600–18 900	3	24 100–25 400	8
18 900–20 200	4	25 400–26 700	9
20 200–21 500	5	26 700–28 000	10

Table 3.3 Projected populations at the end of the design life for the towns used in Table 3.1

Town	Projected population	Score	Town	Projected population	Score
A	27 500	10	G	17 000	2
B	26 500	9	H	16 500	2
C	22 000	6	J	16 000	1
D	18 500	3	K	16 000	1
E	17 000	2	L	15 500	1
F	17 000	2			

nity populations in towns 'A' to 'L' in Table 3.1. Table 3.3 shows individual community scores for projected population at the end of the design life. The ways that the other scoring ranges were determined are shown in Tables 3.4–3.9.

3.3.2 Weighting

Although all the criteria included should be relevant to the selection process some will be more important than others. Applying a weighting system is a way of reflecting the relevant importance

Table 3.4 On-site sanitation failure

Description	Score
On-site sanitation has already failed	10
On-site sanitation is very likely to fail during the design life	7
On-site sanitation may fail during the design life	4
On-site sanitation is unlikely to fail	1

Table 3.5 Polluting industries

Description	Score
More than 50% of the effluent is generated by industry	10
25–50% of effluent is generated by industry	5
Less than 25[1] of waste effluent is generated by industrial sources	1

Note: As these guidelines relate only to sewerage the effect of industrial effluent strength on treatment costs has been ignored.

Table 3.6 Unit cost of construction

Unit cost	Score	Unit cost	Score
40 000–45 000	10	65 000–70000	5
45 000–50 000	9	70 000–75 000	4
50 000–55 000	8	75 000–80 000	3
55 000–60 000	7	80 000–85 000	2
60 000–65 000	6	85 000–90 000	1

Table 3.7 Tourist impact

Description	Score
Over 80% of the community's income is generated by tourism	10
40–80% of income is from tourism	5
No significant tourism income	1

Table 3.8 Consumers' ability/willingness to pay for sewerage

Description	Score
Community can/will repay the full capital and operating costs of the scheme	10
A 50 per cent subsidy is required on the construction cost	7
The community can only afford/is willing to pay operating costs	4
A subsidy is required to make operating costs affordable	1

of the criteria in the selection process. The problem with weighting systems is that it is difficult to apply them objectively. In Table 3.1 'tourist impact', for example, is shown as three times more important than 'projected population'. This is a

Table 3.9 Unit annual operation and maintenance costs

Description	Score
No pumping is required	10
Part of the sewerage network requires pumping	5
All of the sewerage has to be pumped to the outfall or treatment plant	1

reflection of the importance of tourism to the national economy in which the example given was carried out. However, to say that it is three times more important is purely arbitrary. Criteria weights can only be decided upon after discussions with all groups concerned. As they are so arbitrary it is important that a sensitivity analysis is conducted on the results. This is discussed in more detail in the next section.

3.3.3 Scoring and sensitivity analysis

The total score for a community is the sum of all the weighted scores (as shown in Table 3.1). Communities can be ranked with the highest score being first and the lowest score being last. The high-ranking communities are those most favoured for sewerage. The results of ranking can never be completely objective because

Table 3.10 Community ranking

Rank	Town in Table 3.1	Weighted score	1st re-ranking	Weighted score	2nd re-ranking	Weighted score	Priority
1	B	90	B	93	B	54	
2	J	79	J	84	L	44	High
3	L	77	L	80	C	43	
4	D	66	C	66	J	41	
5	C	65	F	60	F	38	Medium
6	F	58	D	58	D	32	
7	G	39	A	41	A	26	
8	A	38	E&G	37	E&G	23	
9	H	37					Low
10	E	36	H	34	H	22	
11	K	24	K	29	K	17	

of the weighting procedure, but confidence in the results can be increased by carrying out a sensitivity analysis. This is done by looking at the effects on ranking caused by changes in the weighting. Select the criteria most open to subjective interpretation. Change the weights of those criteria slightly, recalculate the final scores and re-rank the communities. Comparisons between the ranking order for different scoring and weighting scenarios will indicate the level of confidence that can be given to the results. Minimal changes in ranking order indicate a high degree of confidence in the results. Conversely, wide variations in ranking will indicate that the results should be treated with care. Table 3.10 shows the effect of reducing the weighting for 'tourist impact' from 3 to 2 and increasing the weight for 'unit cost' of construction from 2 to 3 (first re-ranking) and setting all the weights to unity (second re-ranking). Table 3.10 shows that, with the exception of town 'J', the change in weighting numbers has very little effect on the overall ranking of the communities; therefore, the results can be accepted with confidence. The actual numbers are of little relevance, only their relative size.

3.4 CONCLUSIONS

A matrix is an excellent method for deciding the order in which communities should be provided with sewerage. By selecting communities using criteria based on need, benefit and cost, the steps in the process can be justified and the final results can be described mathematically. The method is not absolute. It will not select the single most needy community for sewerage. It will, however, segregate a group of communities into those which can cost-effectively obtain the greatest benefit from sewerage and those with less need or requiring greater subsidy. There will normally be a group in the middle (as has occurred in the example) for which the costs and benefits from sewerage make its provision optional. The individual ranking within that group is irrelevant.

The selection process described here is primarily technical and financial. Environmental and health considerations also have a place in the selection process. These could be included in the matrix or considered separately. Reference may be made to World Bank (1991) for further details of these areas.

3.5 REFERENCES

Franceys, R., Pickford, J. and Reed, R. (1992). *A Guide to the Development of On-site Sanitation*. Geneva: World Health Organisation.

Reed, R. (1994). Why pit latrines fail: some environmental factors. *Waterlines*, **13**(2), 5–7.

STWI (1993). *The Sewerage Masterplan for the Islands of Mauritius and Rodrigues—Development Programme—Rural Areas*. Final report prepared for the Ministry of Energy, Water Resources and Postal Service, Governments of Mauritius. Birmingham: Severn Trent Water International.

World Bank (1991) *Environmental Assessment Source Book—Volume II: Sectoral Guidelines*. Technical Paper No. 140. Washington, DC: The World Bank.

4

Low-cost Sewerage Systems in South Asia

Kevin Tayler

4.1 INTRODUCTION

4.1.1 Background

Cities in South Asia are growing rapidly. In 1992, the urban population of the subcontinent was almost 300 million, of which about 110 million lived in cities with populations of over 1 million, and was growing at about 10 million people per year (World Bank, 1994). India alone had an urban population of about 230 million. Rapid population growth creates a need for shelter and this need is largely met through the informal sector, that sector of the market which operates outside official government rules and regulations. A recent study in Pakistan estimated that the informal sector accounts for about 40 per cent of all new housing and that 49 per cent of the population of the Punjab, the largest province, live in informal areas. The figures for other countries in the region are likely to be similar.

Informal land subdivisions vary greatly in scale and character, from large regularly planned developments which exhibit many of the characteristics of formally approved settlements, to the so called 'slum' settlements which are found on marginal land in and around many Indian cities. They are rarely provided with services at the time they are laid out and the subsequent provision of services places severe strains on the limited resources of government. Particular problems are often faced in relation to sanitation and drainage and it is with possible responses to these problems that this chapter is concerned.

4.1.2 Sanitation options

The available sanitation options can be divided into broad categories for planning purposes. The basic choices are between wet and dry systems and between those which retain faecal material on or near the plot and those which remove it from the plot. Until recently the common practice in most low-income urban areas in South Asia was the use of dry latrines which sometimes incorporate containers which turn them into crude bucket latrines (Pickford, 1983). Such latrines present obvious health hazards. In recent years, there has been a general move towards the use of pour-flush latrines which, while improving on-plot sanitary conditions, also introduce the need for satisfactory effluent disposal. Disposal of latrine effluents to leach pits or soakaways, situated on or adjacent to plots, is possible where plot sizes are not too small and this option has the advantage that the effluent is contained on site and so does not pollute receiving water courses. The double pit design has been widely promoted in India and elsewhere, but it requires more of the householder than other options; Cotton and Franceys (1987) suggest that users do not always understand the principles upon which it is based. Experience with a scheme which preceded those in the Peshawar Cantonment which are described later in this chapter supports this view. Many householders have concreted over the coverslabs of the pits together with, in at least one case, with the flow division chamber. A disadvantage of all on-plot options is that they can only deal with very limited amounts of sullage water. In the Baldia area in Karachi, Pakistan, for instance, a project to dispose of WC effluents to leach pits, although initially successful, experienced some difficulties when the water supply was improved so that sullage water began to collect in streets and a rise in the water table resulted in flooding of the leach pits (Pickford, 1990).

In low-income areas in Pakistan, householders commonly connect WCs to individual septic tanks located in the lane or street outside their houses from which effluent is discharged to open drains. While this practice is better than that of discharging effluents directly to open drains, there must be questions about the fact that the effluent is highly contaminated with pathogens, even after passing through the septic tanks. Sewers would appear to offer a better option than open drains and the experiences of the Orangi Pilot Project (OPP) and others suggest that sewers can offer a technically viable method of waste disposal which is affordable to low-income users. However, Orangi is located in a hilly area and most sewers can be laid to good slopes to discharge to the gullies or nalas, which form its natural drainage routes. Questions arise as to the suitability of sewers in other situations. This chapter explores the extent to which these questions can be answered with our present knowledge, drawing on the experience of projects in Pakistan and India. Some of these involve an approach which is essentially conventional, although incorporating some innovative features, while others involve the use of interceptor tanks on house connections. These are referred to throughout the chapter as conventional and sewered interceptor tank systems respectively. The emphasis is on systems to deal with wastewater although almost all sewerage systems carry some storm flow on occasion.

4.1.3 Project descriptions

This chapter describes experience with four projects in Pakistan and two in India. Of the projects in Pakistan, two are in the government sector and two in the non-government sector, while the Indian examples form part of a World Bank-funded urban development project in Tamil Nadu. Each of the examples is now briefly introduced and its relevance to this chapter is explained.

Yusufabad Housing Colony

Yusufabad is an area of new housing just outside Peshawar, Pakistan, which has been developed since 1986 by the St Michael's Housing Society, a Christian housing society. Many of the members of the society are on low incomes and so from the beginning every effort has been made to keep costs within affordable limits. At present, the scheme comprises about 300 plots, most of which are a standard 98 m² in size, although it has grown over the years. Rights of way are generally 3 or 5 m. Water is piped to all plots from an artesian well and most plots are provided with a core consisting of a single room, a WC/bathroom block and a boundary wall at the time that they are laid out. It was decided at an early stage that conventional sewerage would be the most appropriate option for the disposal of foul water and the scheme was used as a testing ground for some of the ideas described in this chapter. The project was implemented using the services of trainees at a technical training centre funded by GTZ with technical assistance provided by the author.

North East Lahore Upgrading Project

Ths project is a World Bank-funded project which covers an area of about 270 ha on the outskirts of the city. Preliminary planning work on the project began in 1986 and implementation began in 1988 and was programmed to continue until 1994. The project area includes some older settlements, including one village with a history of almost 1000 years, but most of it has been developed over the last 20 years or so. Much of the area comprises informal residential settlements but there is also a considerable amount of industry and commerce. Rights of way are generally narrow with most residential lanes being about 3 m in width. Through routes, although narrow, often carry heavy traffic from factories and workshops. The project was implemented by a special cell formed within the Metropolitan Planning Wing of the Lahore Development Authority (LDA). The author's firm, Gilmore Hankey Kirke, provided technical assistance in the early years of the project, but in 1988 the role of consultants was expanded to include all responsibility for detailed design and the preparation of contract documents and an advisory role on implementation. Many of the standards and procedures referred to in this chapter were established in the course of this project.

Sanitation improvements in Peshawar Cantonment Bustis

This project grew out of the work with the St Michael's Housing Society. Many sweepers and other people in low-status occupations who work for the Cantonment Board live in settlements or bustis ranging in size from about 20 to over 100 households. These settlements are based on rows of single-room dwellings which were originally built as stables but which were later converted to residential use with the provision of minimal shared water supply and sanitation facilities. Residents, who rent the basic accommodation from the Cantonment Board, have added 'katcha' (mud-walled) extensions over the years. At the time that the project was conceived, the water supply to many of the bustis was poor and most of the shared sanitation facilities were poorly maintained and completely insanitary. Various sanitation options were considered but in the end a sewered interceptor tank system was used for most of the bustis. The work was supported by grants from Catholic Relief Services (CRS) in Islamabad, but a condition for proceeding with any scheme was that its intended beneficiaries should raise 10 per cent of its cost. The project was implemented without any ongoing engineering support, the main 'actors' being a Canadian with little previous relevant experience and an Afghan foreman with general building skills but no previous experience of sewer construction. It provides an example of the use of the interceptor tanks in situations in which conventional sewerage would have been impossible and illustrates the fact that interceptor tank systems are well suited to situations in which limited technical resources are available.

Upgrading in Kulakarai, Madras

Kulakarai is a hutment settlement which was chosen as a pilot for the implementation of interceptor tank systems in similar areas. The work formed part of the World Bank-funded Tamil Nadu Urban Development Project (TNUDP). The settlement housed about 110 families on a horseshoe-shaped site which encircled three sides of a pond, the fourth side of which was bounded by a road along which a collector sewer already ran. The work was carried out by the Tamil Nadu Slum Clearance Board (TNSCB).

Thirukatchur sites and services scheme, Madras

Like the Kulakarai scheme, that at Thirukatchur falls under the umbrella of the TNUDP. Thirukatchur lies some 30 km out of Madras on the road to Madurai and the scheme was implemented by the Tamil Nadu Housing Board (TNHB). It is one of many sites and services schemes implemented under the TNUDP and, like the Kulakarai scheme, was viewed as an experiment, once again for sewered interceptor tanks, although the tanks were rather larger than those at Kulakarai.

Faisalabad Area Upgrading Project (FAUP)

The FAUP is an integrated upgrading project with a rather wider scope and a very different approach from the North-east Lahore project. It is participatory in nature, giving a central role in the identification and implementation of project activities to people living in the areas to be upgraded. This emphasis on participation owes much to the fact that prime responsibility for the project within the Overseas Development Administration (ODA), the external donor, rests with the Social Development Department. In view of the innovative nature of the project, it was decided from an early stage that a process approach should be adopted in order to provide the flexibility required to respond positively to community needs and demands. At present, the project is operating in four pilot areas, each of which has a population of about 15 000. Sanitation and drainage have been early concerns of community members in the pilot areas and several small sewer projects have been implemented in one area. They are interesting both for their technical aspects and for the insights that they provide on community involvement in sewer construction.

4.2 STANDARDS

The common assumption that sewerage is unaffordable to low-income users is challenged by the Orangi experience. OPP itself emphasizes the part of community management in eliminating kick-backs to contractors and improving standards of workmanship. However, another important factor is its adoption of appropriate standards. Government agencies in South Asia and in most developing countries use design and construction standards which are modelled on those used in the industrial countries of the North. There is much talk, not least within international agencies such as the World Bank, of the need to reduce standards to levels which are affordable to low-income people. Over the years, efforts to reduce standards have been resisted by government agencies in many countries, largely it seems because they equate reduced standards with second-rate standards. A more promising approach is to start from the premise that standards should be appropriate, recognizing that *they should be framed in relation to the location of a facility and the function which it is intended to perform.* The consequences for·various aspects of sewerage are discussed below.

4.2.1   Minimum sewer depth

Shallow sewers, expecially those which are less than about 1.2 m deep, can be significantly cheaper than deeper sewers. Where labour costs are low, this is not primarily because of any saving in excavation costs but rather because access to shallow sewers for cleaning and maintenance purposes can be gained through inspection chambers which do not require man entry and can thus be much smaller than conventional manholes. The depth of a sewer at any point along its length is affected by its depth at its head and the subsequent gradients. Thus the minimum allowable sewer depth and minimum sewer gradients are important determinants of the cost of a sewerage scheme

All the projects from Pakistan described in this chapter, apart from the Peshawar Cantonment Busti schemes use locally produced concrete sewer pipes. At the time that proposals for Yusufabad were being developed, it was clear that these pipes could take fairly high loads. At one point on the access road to the scheme, a 300 mm pipe was used as a culvert to carry wastewater from an existing artesian installation under the road. Although the cover on this pipe was minimal, not more than about 150 mm, and the road was unsurfaced, the pipe had suffered no damage from occasional traffic, including some loaded trucks. In view of this, it seemed reasonable to assume that, from a structural point of view, pipes could be laid with minimal cover in the narrow lanes of the housing scheme which would not be accessible to heavy traffic.

The other factor was the depth required to ensure that house connections could be made to the sewer at a reasonable slope. The toilet and bathroom facilities of the core housing units were located at the front of the plots and it was found that gradients of about 1 in 40 could be maintained on connections if the invert level of the main sewer was at least 500 mm below ground level. Some reduction in invert depths would be possible if the floor levels of sanitation blocks were raised more than about 150 mm above the lane level but 500 mm depth to sewer invert was adopted as the minimum standard for the scheme. The same standard was adopted for the North-east Lahore project for lanes of 3 m width and under on the basis that vehicle loads would not be a problem and that surveys had shown that most sanitation facilities were located close to the front of plots. Lanes in Faisalabad are generally rather wider, typically 5 m or more, and are accessible to some vehicles. There has been some debate on what the minimum cover should be in these circumstances but the evidence suggests that pipes can take at least moderate traffic loads at covers of about 250 mm. The standards given in Table 4.1 are those suggested by Tayler and Cotton (1993) based on the North-east Lahore experience.

The sewers in the Peshawar Cantonment schemes are laid in narrow lanes, typically less than 1.5 metres wide, where there is no possibility of any traffic loading. In these lanes, locally made plastic pipes have been laid with covers of about 150 mm. More research is needed to determine standards for the cover to be provided over plastic pipes in locations were vehicle loads can be expected.

Clay pipes are not used in Pakistan but are not uncommon in India. A British clay pipe manufacturer (Naylor Bros (Claywares) Ltd, Barnsley. 1982) suggests the minimum covers shown in Table 4.2. Extra-strength pipes are supplied as standard in the UK but caution is needed as the pipes used in India are not of the same quality as those produced in Britain. As for locally

Table 4.1

Street width	Heaviest vehicle	Minimum cover
< 3 m	Motorcycle	250 mm
3–4.5 m	Suzuki car or van	350 mm
4.5–6 m	Cars, horse drawn carts, small trucks	400 mm
> 6 m residential	Occasional trucks	500 mm

Table 4.2

Pipe diameter		100 mm diameter	150 mm diameter
Minor roads	Extra strength	400 mm	600 mm
	Super strength	350 mm	450 mm
Gardens	Extra strength	350 mm	350 mm
	Super strength	350 mm	350 mm

produced plastic pipes in Pakistan, research is needed to determine standards for the covers to be provided over locally produced clay pipes. The minimum covers specified in Thirukatchur were about 700 mm.

4.2.2 Minimum sewer gradient

In flat areas, the minimum sewer gradient has possibly more influence on the depth of a sewer throughout its length than the minimum permissible cover. Conventional theory is not suitable for the design of sewers near the head of the sewer system as the steady-state flow conditions which it assumes do not occur in such sewers. British codes overcome this problem by providing rules of thumb for such sewers. The building drainage code BS8301 (British Standards Institute, 1985), for instance, recommends the following minimum gradients:

100 mm diameter sewer: 1 in 40 if peak flow less than 1 l/s; 1 in 80 if peak flow more than 1 l/s
150 mm diameter sewer: 1 in 150 if at least five WCs are connected.

Sewers for Adoption (WSA, 1995) recommends a gradient of 1 in 150 for sewers receiving flows from at least 10 dwelling units but this, as with the BS8301 recommendations appears to be a rule of thumb.

Various attempts have been made, mostly applying probability theory to the likely timing of discharges and using information on the flow produced when a WC is flushed, to develop a rational design theory for sewers which are subject to unsteady flow conditions. Most of these are based on European or North American conditions but at least one (UNCHS, 1986) is based on research in Brazil. This argues for a minimum gradient of 1 in 167, based on the assumption of a flushing rate of 2.2 l/s. The latter is high for a pour-flush system and ignores the effect of flow attenuation. An important practical point which has emerged in several of the projects reviewed in this chapter is that house connections often enter manholes at some distance above the invert level of the sewer. This is a departure from conventional practice which requires that branches enter the main sewer with their soffits at or slightly above that of the main sewer, if necessary following a backdrop pipe, so that benching can be provided to guide flows and conserve momentum. It is likely to invalidate some of the assumptions which are made in the various theoretical attempts to arrive at a rational design philosophy.

The gradients adopted in North-east Lahore were related to the number of houses connections and have been reported previously (Tayler, 1990). The aim was generally to achieve a gradient of about 1 in 120 at the head of a sewer with progressively flatter gradients allowed as the number of house connections increased. The minimum gradients adopted for tertiary sewers, receiving flows from up to about 100 houses, were about 1 in 180. In a more recent publication (Tayler and Cotton, 1993), slightly flatter gradients were suggested in accordance with Table 4.3. These figures are to some extent empirical. They start from the assumption that a gradient of 1 in 150 is acceptable for a sewer receiving flow from 10 houses and extrapolate between this assumption and a conventional approach to the design of a 230 mm diameter sewer running full (in the latter case, the assump-

Table 4.3

No. of houses	Sewer gradient
10	1:150
20	1:162
40	1:175
60	1:180
100	1:190
200	1:200

tion is made that the sewer carries a flow of 6 x dry weather flow when running full). Analysis suggests that, with a k_s value of 6.0 in the Colebrook–White equation, these gradients give maximum dry weather flow velocities ranging from about 0.45 m/s for 10 houses, through about 0.5 m/s for 40 houses and 0.55 m/s for 100 houses to 0.6 m/s for 200 houses. These velocities are lower than those which are normally specified but they assume a more realistic k_s value as locally made pipes may have a higher friction coefficient than normally assumed (the internal surfaces of Indian and Egyptian clay pipes are often decidedly rough, whereas Pakistani spun concrete pipes are smoother but those produced by small informal sector casting yards often have a rather uneven finish).

Studies in the UK (Lillywhite and Webster, 1979) suggest that gradient is not the main determinant of sewer blockages, which are affected more by the condition of the sewer and in particular of its joints. Poorly made joints are likely to cause blockages. However, the relevance of these studies, which were presumably conducted in relatively clean lengths of sewer, to the conditions found in South Asian sewers is debatable. Field observations over a number of years suggest that most conventional systems in Pakistan are obstructed to a greater or lesser extent, creating flow conditions which are certainly not self-cleansing and would be extremely difficult to represent theoretically. Solid waste and debris are common causes of obstruction, but another important factor is a build up of silt and sand in the inverts of sewers over time. Some of the sand may be washed into sewers from kitchens where it is used to clean utensils, but it is probable that the majority of the material is silt washed into the sewer with rainwater. Consideration of all these points leads to the conclusion that it is probably impossible to design sewers in low-income urban areas which are both maintenance-free and affordable. Some compromise between these two objectives will always be necessary.

This is perhaps a subject for further study but one response is to conclude that sewers should be designed so that the settlement of solids is concentrated at locations from which they can relatively easily be removed. Interceptor tank systems, designed to remove solids before flows enter the sewer, provide one means of achieving this objective. Another possible approach, which has been observed in both community and municipality built sewers in Pakistan, is to construct chambers with no benching and with their floor level some way below the invert of the sewer so that they act as interceptors within the sewer itself. This practice is not in accordance with conventional thinking, but one sewer in Lahore to which it was applied was found to be operating satisfactorily over 2 years after construction even though its gradient was less than 1 in 800. Experiments with the practice conducted in Faisalabad are described later in this chapter.

The available literature suggests some variation in thinking on the gradients which are allowable with interceptor tank systems. At one extreme (Otis and Mara, 1985), no limits are given on the minimum gradient; indeed it is argued that some sections can be laid below the hydraulic grade line, implying that some backfall sections are acceptable. The only stipulations are that there should be some overall fall across the system and that the hydraulic grade line during estimated peak flows should not rise above the invert of the outlet from any interceptor tank. Other recommendations are more conservative, reflecting concern that interceptor tanks will not be desludged, so that solids will eventually find their way into sewers. In a previous publication (Tayler and Cotton, 1993), gradients ranging from 1 in 150 for a 75 mm diameter sewer serving about 15 houses to 1 in 450 for a 225 mm diameter sewer serving up to 260 houses were suggested. The Peshawar Cantonment Busti systems have certainly been laid to much flatter slopes, although precise measurements of these slopes are not available. However, questions remain as to the long-term performance of such systems if desludging of interceptor tanks is neglected.

4.2.3 Manhole/chamber spacing

The manhole and chamber spacing adopted for the Yusufabad and North-east Lahore schemes was limited by the fact that the normal practice in Pakistan is to make all house connections directly into manholes. Indeed, this is required by most government departments and agencies which deal with sewerage. There are practical reasons for this. It is much harder to produce good quality concrete branch fittings than it is to produce straight spun concrete pipes and poor quality Y and T branches in a sewer line are potential causes of blockages. In the Yusufabad scheme, each chamber receives the flow from four houses and their spacing is typically about 16 m. Greater spacings were possible in the North-east Lahore scheme because of the approach taken to house connections, which will be described later. The maximum spacing adopted for tertiary sewers was 30 m, although spacings of up to about 100 m were used on collector sewers. The 30 m figure is considerably less than that which would normally be permissible in the UK where the minimum recommended spacings for manholes and inspection chambers are 90 m and 45 m respectively (BRE, 1984). However, the equipment available for clearing blockages is not of the same quality as that in the UK, consisting as it does of lengths of bamboo crudely wired together. Experiments in North-east Lahore proved that lengths of at least 25 m can be rodded with wired bamboo rods, but this appeared to be close to the limit of what was realistically possible.

Observation suggests that many sewer blockages in South Asia are caused by materials deposited into sewers at chambers and manholes, including pieces of broken cover. One report on Karachi (Balfours and Engineering Consultants, 1987) recommends that chambers should not be provided on shallow sewers on the grounds that the reduced possibility of blockages thus achieved more than compensates for the added inconvenience of having to dig down to the sewer and break into it to locate any blockage which does occur. This recommendation assumes that interceptor tanks are provided on house connections. This approach was independently adopted for the Peshawar Cantonment Busti schemes and blockages do not appear to have been a problem with these schemes in the first 2–3 years of their operation. However, they have not been systematically monitored.

Chambers have been omitted from some short privately built conventional sewers in Pakistan but the extension of the approach to public sector sewers is problematic because it is contrary to conventional wisdom and existing codes and regulations. There is, in any case, the problem of producing the special pipe fittings required to make connections at locations other than chambers. One possible compromise solution would be to make connections at chambers with covers located below ground level, as shown in Figure 4.1. Apart from reducing the opportunity for casual access to the sewer, this arrangement can also reduce costs by allowing the use of a simple unframed cover and should also

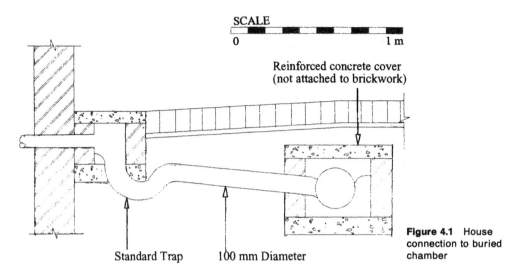

SCALE

0   1 m

Reinforced concrete cover
(not attached to brickwork)

Standard Trap   100 mm Diameter

Figure 4.1 House
connection to buried
chamber

reduce the likelihood of the cover being damaged by traffic. It is
significant that residents in both Faisalabad and Karachi have
been observed to bury conventional chamber covers in order to
protect them from traffic.

4.2.4 Dimensions of access manholes and chambers

The function of manholes and chambers is to allow access to the
sewer in order to inspect it and clear any blockages which may
occur in it. A manhole, as its name suggests, has to be entered in
order to carry out these operations, while a chamber can be
accessed from ground level and requires rather smaller plan
dimensions.

The internal dimensions adopted for access chambers in Yusufabad were 500 × 500 mm. These dimensions are adequate when
the sewer is very shallow but it is doubtful whether it would be
possible to rod the sewer from such chambers when the depth to
invert exceeds about 800 mm. Bearing this in mind, Table 4.4
gives standards based on those adopted for the manholes and
chambers used in North-east Lahore (the dimensions given in
bold type are those actually used while those in normal type are
roughly equivalent alternatives).

A comparison may be made with the standards recommended
in *Sewers for Adoption* (WSA, 1995) and with the recommenda-

Table 4.4

Depth to sewer invert (mm)	Entry required	Rectangular manhole/chamber dimensions (mm)	Circular manhole/chamber diameter (mm)
>800	No	**600 × 500**	525
800–1350	Partial (standing on benching)	**800 × 600**	750
1350–2000	Yes	1200 × 750	**1050**

Table 4.5

Depth to sewer invert (mm)	Rectangular or circular chamber	Dimensions (mm)
<1000	Rectangular	900 × 600
1000–1350	Rectangular	1200 × 675
1350–1500	Circular	1050 diameter
1500–3000	Circular	1200 diameter

tions of the Building Research Establishment (BRE, 1984). The recommendations contained in *Sewers for Adoption* are in Table 4.5.

The diameters for circular manholes are for sewers of 300 mm diameter and less. The BRE recommendations are for building drainage and are similar to those used in North-east Lahore—indeed they suggest that 190 mm diameter chambers are appropriate for invert depths of 600 mm and less. BRE's recommendations for greater invert depths are given in Table 4.6.

One of the most difficult tasks in developing cost-effective approaches to the provision of sewerage within the public sector

Table 4.6

Depth to sewer invert (mm)	Internal dimensions (mm)	Nominal cover diameter (mm)
<1000	450 × 450 or 450 diameter	450 × 450 or 450 diameter
1000–1500	1200 × 750 or 1050 diameter	600 × 600 or 600 diameter

is to convince engineers to accept changes from universally applied 'conventional' standards. The figures in Table 4.6 show that the North-east Lahore manhole dimension standards were comparable with those used in similar situations in Britain. Some caution is required in view of the different sewer cleaning equipment available in the two countries. Purpose-made flexible drainage rods are available in Britain and in many cases it is possible to call upon high-pressure jetting equipment which can be much more effective than drainage rods in removing blockages. In Pakistan, as already indicated, the normal practice for small diameter sewers is to use lengths of bamboo, wired together on site in a rather crude way. One of the disadvantages of this practice is that the wired bamboo rods are likely to be much less flexible than purpose-made rods with screwed connections, thus requiring rather larger chambers to allow them to be bent and inserted into the sewer. The response to this situation should be to put more effort into the development of more effective methods and higher-quality equipment for clearing sewer blockages and this is certainly something that should be given priority in future externally funded schemes. However, it is likely to be many years before such improved methods and equipment become widely available and there is a need for more experimental work to establish the limits of what can be achieved with presently available equipment.

All manholes and chambers in Yusufabad and North-east Lahore had brick walls and in view of this it was felt that a rectangular shape was preferable for the smaller chambers. The 600 × 500 mm chamber is provided with a single cover slab while a split cover is required for the 800 × 600 mm chamber. The FAUP team has experimented with the use of circular concrete chambers using shuttering based on that developed by the OPP and it has been found that reasonable quality can be obtained using this method for 525 mm diameter chambers. The 750 mm diameter chamber has yet to be tested. In order to allow easy access, the removable cover should extend over the full plan area of a chamber and there may be problems with the cover on a 750 mm diameter chamber which, at about 900 mm in diameter and 75 mm deep, will weigh about 110 kg.

The normal practice for manholes designed for entry is to provide a fixed cover slab with a smaller opening, typically about 550 mm in diameter, through which the manhole can be entered.

Problems arise with this arrangement when the depth to invert is less than about 1700 mm because the height available between the benching and the cover slab provides insufficient working space. There appears to be no entirely satisfactory way of dealing with this problem, but arguably the best response is to accept that chambers should be used for rather greater depths to invert than suggested above, perhaps to about 1500 mm.

4.2.5 Minimum sewer diameter

The flows which occur in tertiary sewers are low and rarely exceed 2–3 l/s. As the capacity of a 100 mm diameter sewer laid at a gradient of 1 in 100 is about 5 l/s, the flow to be carried is rarely the factor which determines the size required for such sewers. Most water and sewerage agencies in Pakistan specify a minimum sewer diameter of 9 inches (230 mm). This size is greater than that which is suggested by hydraulic theory and provides less favourable flow conditions than would occur in a smaller pipe. However, attempts to reduce pipe diameters are normally resisted on the grounds that, as some settlement of solids is bound to occur, the larger pipe diameter will enable the system to function even when it is partially blocked. This attitude is not confined to agency officials but is also widespread among community members. In Yusufabad, some short lengths of sewer, serving up to about 20 houses, are 150 mm diameter but residents definitely prefer to use 230 mm diameter pipes. The minimum diameter adopted in North-east Lahore was 230 mm. In Faisalabad, the first 100 m or so of the first conventional sewer constructed through the FAUP was 150 mm diameter and discussions are continuing as to whether some use of 150 mm diameter sewers will be acceptable to the government authorities. To date, after about 9 months use, no problems have been reported with this length of sewer.

The Peshawar Cantonment Busti interceptor tank schemes use 75 mm pipes to serve up to at least 25 houses. These appear to be working satisfactorily several years after construction, although no systematic evaluation has been carried out. The one interceptor tank sewer that has been constructed in Faisalabad is 150 mm diameter, reflecting the fact that government authorities and community members tend to be conservative on the subject.

One hundred mm diameter sewers were used with the interceptor tank systems in Kulakarai.

4.2.6   Dimensions of interceptor tank chambers

Information is available on the interceptor tanks incorporated into three of the schemes considered in this chapter, those in the Peshawar Cantonment, Kulakarai and Thirukatchur. Those used in the Peshawar schemes are circular in plan with an internal diameter of 0.9 m and a depth of just over 2 m. The design was based on the perception that there would be limited maintenance and that there would be a reluctance to desludge the tanks at regular intervals. The size would provide sufficient capacity for tanks to work for several years (calculations suggest between 4 and 5 years) before desludging became necessary and it is as yet too early to say what will happen when they are full of sludge. However, it is interesting to note that communal septic tanks, receiving flows from the WCs of up to about 20 houses, which exist in the Civil Lines area of Peshawar Cantonment, are possibly 50 years or more old and are still operating, although it does not appear that they have ever been desludged. These septic tanks discharge to open drains and it is likely that they no longer perform as designed, but it may be that they are still carrying out the main function of an interceptor tank, to remove gross solids from the sewer.

In complete contrast, the tanks used in Kulakarai were a mere 800 × 600 mm in plan with a total depth of 650 mm. Of this depth, only about 350 mm was below the invert level of the outlet pipe, giving only about 6 months sludge retention. When the scheme was inspected, some 9 months after it had been commissioned, it was working satisfactorily, but it would be interesting to carry out a follow-up investigation now, some 5 years after commissioning. The tanks on the Thirukatchur scheme were 1000 × 600 mm in plan and about 650 mm deep below the invert of the outlet pipe. The available sludge storage volume is thus about 0.39 m^3, intermediate between those in Peshawar and Kulakarai.

One of the sewers implemented in Faisalabad is also designed to include interceptor tanks which are similar to the septic tanks commonly provided in Pakistan to provide some treatment prior to the discharge of WC wastes to open drains. The dimensions of these are typically similar to those at Thirukatchur.

Observation of the operation of household septic tanks discharging to open drains, which are not normally fitted with a sanitary tee on the outlet, reveals that finely divided solids are washed out every time a flush of water enters a tank. Bearing in mind the fact that the main purpose of interceptor tanks is to remove gross solids rather than to provide treatment, it is arguable that sanitary tees should be omitted from them, allowing material from the floating crust in the tank to be washed into the sewer and thus reducing the rate of sludge accumulation. The material flushed out should not cause problems in sewers because of its low-specific gravity and finely divided nature.

4.2.7 House connections

It is commonly assumed that the failure of householders to make connections to sewers is a significant problem in low-income areas. This was perceived to be a potential problem in North East Lahore where there was little community involvement in either planning or implementation and government efforts did not extend beyond plot boundaries. In order to minimize future disruption, it was decided that existing open drains along the sides of lanes should be covered and in effect converted into shared connections. This system did not work well for two reasons: first, because householders continued to view the connections as drains and they did not worry if gaps appeared in the cover slab for any reason, and secondly because no one took responsibility for maintaining the side drains. The water authorities said, rightly, that the drains were not sewers; the municipal authorities said that they were only equipped to clean open drains; and the householders were reluctant to take responsibility for a shared facility which they had not requested. Subsequently, experiments were made with replacing the covered drain arrangement with 100 and 150 mm pipes and these appeared to work better. However, in principle, it seems better to make connections from individual houses to a sewer wherever possible.

A more positive approach to the connection problem is to recognize that the problem is likely to disappear if residents are involved in the planning and implementation process for local sewers. This is certainly the case in Orangi and preliminary evidence suggests that the situation will be similar in Faisalabad. Involvement of residents shifts the interface between the 'private and public' domains from the plot boundary to the point at which local sewers discharge to collector sewers and allows a more integrated approach to be taken to local sanitation and drainage improvements. This more integrated approach should arguably allow schemes to tackle the problem of sand entering sewers through house connections by encouraging residents to provide grit traps on sullage lines. It should also be possible to encourage them to provide traps on connections to deter cock-roaches and rats. However, more field research is required to determine the best connection arrangements and to develop methods to encourage residents to adopt them.

4.3 CONSTRUCTION DETAILS AND PROCEDURES

4.3.1 Pipe materials and jointing

Government agencies in Pakistan specify spun spigot-and-socket concrete pipes with each spigot provided with a groove designed to allow the fitting of a rubber ring. While this should make the pipes flexible, the reality is that little faith is placed in the efficacy of the rubber rings and the normal practice is to finish each joint by filling the annulus between the spigot and the socket with cement mortar. This means, of course, that the joint is rigid rather than flexible. Early in the North-east Lahore project a decision was made to discontinue the use of rubber rings as they served no useful purpose once the decision to mortar the joint had been made. The pipes used in North-east Lahore were produced in one of the larger casting yards which had been approved by government. Pipes of 300 mm diameter and above were produced broadly in line with the requirements of the rele-vant ASTM specification, while 230 mm diameter pipes complied with BS5911.

  In the informal sector, most sewers are constructed using plain-ended concrete pipes which are purchased from small local casting yards which also produce a variety of materials used in house construction and for other purposes. The pipes are produced in rapidly rotating horizontal moulds, using the same process as that used in the larger casting yards. The reinforcement used is nominal and does not comply with the BS and ASTM standards used by the larger casting yards. The internal surface of pipes inspected at a small casting yard in Faisalabad was rather uneven and this might increase the roughness slightly over that normally assumed in hydraulic calculations. Joints are made by inserting a strip of jute soaked in cement slurry into the gap between the pipes and then surrounding the whole joint in sand–cement mortar.

The use of plain-ended pipes is not in accordance with government specifications, but experience in Yusufabad and the FAUP suggests that sewers built in this way perform perfectly satisfactorily. This is significant, given that the price of plain-ended pipes is about one-third of that of those with sockets.

Clay pipes used in India are also of the spigot-and-socket type, as are the locally made plastic pipes and fittings which are available in Pakistan and no doubt elsewhere. Joints in the latter are normally made to allow a push fit and this is perfectly satisfactory for the small diameters required for branch sewers in interceptor tank systems.

4.3.2 Pipe bedding

Sewers in Yusufabad were laid in trenches in previously undisturbed fine-grained soil which provided ideal conditions. Pipes were laid directly on to the trench bottom. The common practice in informal areas in Lahore is to raise house plinth levels some way, typically about 1–1.5 m, above the natural ground level, and it is not uncommon for solid waste to be used to bring lane levels up to the required height. Thus the ground condidtions are rather variable and unreliable. All sewers in the North-east Lahore scheme were laid on a 100 mm brick ballast bed, brick being used because the nearest source of stone was over 100 km away. Bedding is required by Pakistan Government specifications, as it was for the World Bank-funded schemes in India. The early community-managed sewer projects implemented under the FAUP have not included bedding and this is one among several examples of where government procedures and those adopted by local communities are likely to be in conflict.

4.3.3 Manhole construction

In the Yusufabad and North-east Lahore schemes chambers are rectangular, constructed in brick on a concrete base with concrete covers. In North-east Lahore, the walls were originally half brick (112 mm) thick for 500 × 600 mm chambers and single brick thick (230 mm) for 800 × 600 mm chambers. It was found that half-brick thick side walls were unsatisfactory where more than one pipe entered a chamber along one of its sides and the design was therefore amended to allow a full brick thickness where this was the case. In both schemes, the covers were cast in an angle iron frame and located within a second frame, which was built into the surround at the top of the manhole. In North-east Lahore, some of these frames were galvanized, but the galvanizing was not found to be satisfactory and painting of the frames in bituminous paint was then adopted as standard.

In Faisalabad chambers on all but the first sewerage scheme implemented have been constructed using the methods developed by OPP. The base is laid in mass concrete, as for the other designs discussed, but the plan shape is circular and the walls are constructed in mass concrete, placed *in-situ* between steel shutters with box-outs left for sewers and connections. The wall thickness is 100 mm and given the fact that the cost of 1:2:4 concrete is roughly the same as that of brickwork in the Punjab, this gives some cost saving over conventional single-brick thick walls. The arrangement is shown in Figure 4.2.

The breakdown of tender prices for the small chambers used in North-east Lahore showed that the covers and their frame were priced at about twice as much as the rest of the chamber, about Rs1200 against Rs600 for the 800 × 600 mm chambers and about Rs600 against Rs300 for the 600 × 500 mm chamber. (1990 prices; US$1 approximately equivalent to Rs30.) The angle iron frames were the main reason for the high prices of covers. Most manhole covers provided by private individuals and community groups in Pakistan do not have frames, and OPP argue that this is quite satisfactory for their circular covers, although unframed rectangular covers tend to break at the corners. The OPP practice was at first followed in Faisalabad but it was found that even circular covers are susceptible to breakage when subjected to wheel loads. A modified design was therefore developed which included a vertical steel strip encircling the

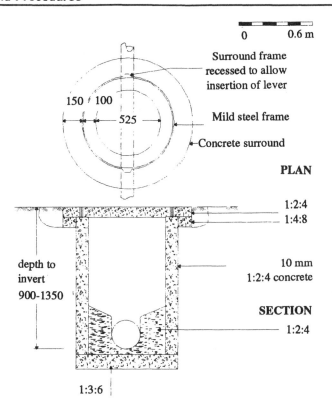

0   0.6 m

Surround frame
recessed to allow
insertion of lever

150 100

525

Mild steel frame

Concrete surround

PLAN

1:2:4
1:4:8

depth to
invert
900-1350

10 mm
1:2:4 concrete

SECTION
1:2:4

Figure 4.2 Manhole used in the Faisalabad Area Upgrading Project

1:3:6

cover and welded to the reinforcement. This was considerably cheaper than the angle iron design and lent itself to use with circular covers.

A weakness of the standard OPP design is that no surround is provided to contain the cover in place. The experience in the early projects in Faisalabad is that this allows the cover to move laterally so that part of it can extend beyond the manhole walls and be subjected to large stresses when a load passses over it. To prevent this, a mass concrete surround will be provided to contain the cover on future projects. It is probably advisable that, the vertical face in contact with the cover should be provided with a steel edge strip as for the cover itself.

The larger circular manholes used in North-east Lahore have single brick walls on a plain concrete base. The cover is a standard Lahore Water and Sanitation Agency (WASA) design: circular in plan and cast in concrete inside a cast-iron frame. The cover is set over a circular access hole in a flat reinforced concrete roof slab, thus conforming with standard sewerage prac-

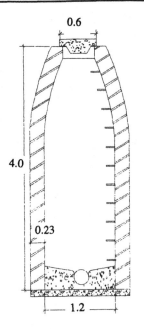
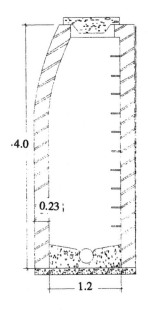

Figure 4.3 Deep manholes in Pakistan: standard design (*left*) and a recommended modification (*right*)

tice. This is a departure from the normal practice in Pakistan which is to corbel in the upper manhole walls, as shown in Figure 4.3, so as to allow a cover to be fitted without a reinforced roof slab. A similar detail is shown on standard drawings of the Maharastra Housing and Development Authority (MHADA) and is probably widespread in India. While it is economical from a construction point of view, it appears not to meet the criterion that standards should relate to practice as sewer workers need to be accomplished rock climbers to descend the step irons which disappear out of view below the overhang. At first sight, it would appear that a better option would be to keep one side of the manhole vertical, corbelling the other side of the manhole over as shown in Figure 4.3. However, investigation suggests that the normal practice in Faisalabad is to lower a sewer man on a rope rather than to use the step-irons, which are often badly corroded. It is arguable that this practice, while far from ideal, is the best option available in the circumstances.

No step irons have been provided in any of the smaller chambers provided in the Pakistani schemes. A previous design recommendation (Tayler and Cotton, 1993) suggested that the plan dimensions of chambers intended for depths to invert between 1000 and 1350 mm should be increased to 1000 × 600 mm to allow for the provision of step irons at one end of the chamber. However, it is interesting to note that most water authorities in England now discourage the use of step irons because they are potentially dangerous when corroded. Given the fact that inspection shows that many step irons on existing manholes in South Asia are badly corroded, this philosophy has much to be said for it, provided, that ladders of a suitable quality are available as an alternative.

Benching was provided in North-east Lahore in accordance with conventional practice. There is a tendency for local government and community schemes to omit benching or provide it only up to the mid-barrel level of the sewer. Chambers in the early schemes in Faisalabad were constructed with their base dropped below invert level and the benching omitted, the aim being to ensure that any large solids entering the sewer would settle at a chamber rather than between chambers. After some months of operation, sludge has built up to above invert level on either side of the flow and, although there is no evidence of large solids being deposited in the sewer, it has been decided that conventional benching will be provided in future chambers. In any case, the dropped manhole invert system is only appropriate for shallow chambers which do not have to be entered for maintenance purposes.

The North-east Lahore chambers and manholes were rendered on their internal but not their external faces, on the ground that the sewers were above the water table so that there was no need to prevent the ingress of water.

4.3.4 Setting out procedures

Perhaps the most difficult task associated with the implementation of sewerage projects is the laying of sewer pipes to the correct levels. This has traditionally been done in the West using profiles set a given distance above the invert of the sewer and travellers of the length necessary to fix the bottom level of the trench, the top level of any bedding and the invert level of the pipe. In Pakistan, a rather inferior version of this method is commonly used by government engineers. Measurements are made from a string line, which is stretched between profiles set across the sewer line at intervals. Both of these methods normally require that levels are fixed using a surveyor's level. Community members and contractors normally have neither access to a surveyor's level nor the ability to use it, so there is a need to explore alternative ways of fixing levels. Two methods are in fairly common use in Pakistan, and no doubt elsewhere. The first involves the use of a water level, in effect a hose filled with water, to determine differences in level between two points. The second is to use a carpenter's spirit level, together with a straight edge, to determine the fall over a short length. Both these methods have limitations. The water level system can give misleading results unless care is taken to remove all air bubbles from the hose. It is therefore unsuitable for use by untrained workers. The carpenter's level method works reasonably well when falls are good, say greater than about 1 in 200, but it is not certain that it can provide sufficient accuracy when the available fall is limited.

In North-east Lahore, the normal practice was for engineers to fix levels on the walls of buildings fronting the lane or street along which a sewer was to run. The trench and invert levels were then tranferred to profiles set over the trench as already described. For community-managed schemes and those implemented by small local authorities with limited resources, it should be possible to develop methods for branch sewers which rely on locally available skills with the engineering input perhaps restricted to fixing the level at one point on the sewer. This will depend on standardized rules for the falls to be provided on branch sewers becoming generally known and accepted.

4.3.5 Contract drawings and documentation

More investigation is required on appropriate contract documentation for small sewer projects. The contracts in North-east Lahore, which included services other than sewerage, typically covered 20–30 ha and were for sums in the range Rs15 million to Rs30 million (US$500 000 to US$1 million), were covered by contract documents based on the FIDIC conditions of contract. This represents a large degree of overkill and it is unlikely that either contractors or government officials are aware of much that is in the contract documents. On the other hand, community and municipality contracts are often carried out with an absolute minimum of documentation and this can lead to mistakes, omissions and acceptance of poor workmanship. In North-east Lahore, the approach to drawings was to produce a range of standard details, those for sewerage covering a range of standard manhole and chamber designs, pipe bedding details and typical arrangements showing how the various components might be used with each other. Plans and sections of collector sewers were produced in the conventional way. Full details of tertiary sewers were shown on 1 in 500 plans. The location of each chamber and manhole was shown, its type was specified and its invert and cover levels were given. The length of sewer between manholes and the average gradient between chambers and manholes, to 1 in the nearest whole number, were shown alongside each sewer leg. This system removed the need to produce a section for every sewer. The chamber and manhole cover levels were used to fix the levels for the lane surfacing and were designed to ensure that surface water could always run off. (Generally, the aim was to keep surface water separate from foul water, allowing it to run for considerable distances on the lane and street surface so as to reduce the length of surface water drains required.)

In Faisalabad, the contracts are much smaller and are undertaken by local tradesmen. Standard details are in the process of being prepared and there will also be a need to develop some form of appropriate documentation. GHK International and WEDC are currently collaborating on an ODA-funded research project to investigate what form this documentation might take.

4.4 COMMUNITY INVOLVEMENT IN SEWER PLANNING AND CONSTRUCTION

4.4.1 Why community involvement?

There are four possible reasons for involving comunity members in sewer planning and construction:

- to reduce costs;
- to increase the rate of implementation;
- to improve construction quality and the standard of maintenance; and
- to develop community capacity and organization.

Costs

It has been stated (Mustapha, 1985) that the cost of sewers implemented by community members with technical assistance provided by OPP is about a quarter of that of similar government schemes. The experience in Faisalabad and North-east Lahore suggests that this rather overstates the saving to be made. The average cost of local sewers in North-east Lahore, excluding house connections, was about Rs1070 per household at 1988–89 prices (GHK and WEDC, 1991). This compares with costs per household in the range Rs900–1000 in Faisalabad (1994 prices). However, the cost of connections in North-east Lahore was high, about Rs2500 per household. This is partly because of the system adopted, but it also reflects the inefficiency of connecting all existing discharges to the sewer. This would have been reduced if community management had meant that the sewers and connections could be considered together so that improvments could have been made in on-plot arrangements at the same time that the sewers were constructed.

Another important point is that beneficiaries bear the full capital cost of local sewer provison in Orangi and 50 per cent of the cost in the FAUP. It is usually very difficult to recover the costs of government-financed schemes from beneficiaries. The assumption in North-east Lahore was that the costs would be recovered through increased property taxes, but this assumption was hardly realistic when about 90 per cent of householders in the area were exempt from such taxes and the department responsible for setting taxes was quite separate from that managing the upgrading work. Where land is owned by government, it is possible to make a charge for regularizing tenure, the level of which is set to recover infrastructure upgrading costs. However, such schemes rarely achieve complete cost recovery because of the failure to increase regularization charges in line with inflation and to ensure that all residents go through the regularization procedure. It has been estimated (GHK and WEDC, 1991) that in Pakistan the overall cost recovery from plot regularization rarely exceeds 25% per cent of the total cost of infrastructure provision. Thus community management at the local level can greatly reduce the financial burden on government.

Rate of implementation

Community involvement at the local level should theoretically allow scarce government resources to be concentrated on the provision of primary and secondary facilities. Whether this will be the case in reality depends on a number of factors, not least the ability of projects like the FAUP to provide support to communities in a cost-effective way. To date, the sewers implemented through the FAUP have required considerable inputs from social organizers, and the aim over the next year will be to reduce the need for such inputs. One way in which it is hoped to achieve this is by providing training to masons and small contractors so that they are better able to install local sewers to an acceptable standard with minimal external supervision.

Construction quality and maintenance

The experience of North-east Lahore suggests that maintenance problems are likely to occur with conventional sewers if the users are not involved in the planning and development process. These problems stem partly from the large amount of solid waste which finds its way into the sewers in the absence of adequate collection services. The problem was initially exacerbated in North-east Lahore by the fact that chamber covers were only 50 mm deep and were therefore fairly light. It was therefore relatively easy to lift them in order to deposit rubbish in the sewers, a practice which has been observed on more than one occasion. As a result of this experience, the cover thickness was increased to 75 mm. The use of interceptor tanks is another way of reducing maintenance problems, but it is not possible to completely 'engineer-out' fundamental problems in this way. Experience elsewhere in Pakistan suggests that the quality of government sewerage schemes is often poor.

The early experience in Faisalabad suggests that community members take great pains to ensure that the quality of materials and workmanship on schemes which they pay for themselves are good. Preliminary indications are that they will be prepared to carry out any maintenance required, including where necessary the replacement of manhole covers.

Development of community organizations

The Faisalabad experience suggests that lane sewers are a good vehicle for the development of cooperation at the lane level, and there are already cases where cooperation on sewer schemes has led groups to develop further proposals, both at the lane level and, in conjunction with other lane groups, at the neighbourhood level.

4.4.2 Scope for community involvement

As has already been indicated, the Orangi situation includes some factors which are often not found elsewhere. In particular, the topography means that local sewers are not dependent on the existence of collector sewers. OPP is also unusual in that it is led by an unusually charismatic founder and has a core team of dedicated professionals. Is community involvement in sewerage provision replicable? Various reports (for example, Balfours and Engineering Consultants, 1987; GHK and WEDC, 1991) have described examples in Pakistan of community-built sewers connected to either government collector sewers or open drains. There are many examples of community-built sewers in one of the FAUP pilot areas in Faisalabad. However, some of these sewers are provided with insufficient fall and are poorly constructed so that they block frequently and are generally unsatisfactory. Experiences such as this lead some professionals to assume that sewerage is too complicated to be planned and implemented using community resources. For instance, sewerage was specifically excluded from the range of tasks considered to be suitable using community contracting procedures in the Million Houses Programme in Sri Lanka (Yap 1993). The experience in Faisalabad suggests that this is not true for local (lane level) sewers, provided that there is a sewer or drain at a suitable level to which the local sewer can discharge.

   This leads to the important conclusion that it is possible for community groups to manage the construction of local (tertiary level) sewers with some external technical assistance but that government and its professionals must take the prime responsibility for overall planning and the provision of primary and secondary sewers and treatment facilities. In other words, they must provide the context within which local community initiatives can be effective.

4.4.3 Experience of community involvement

This discussion of experience with community involvement is based largely on that of the FAUP. The FAUP is still at a relatively early stage, but it does provide some insights into factors which influence the success of community managed sewerage schemes.

Understanding of technical issues

Community members do have knowledge and ideas; for instance, the group responsible for the first sewer implemented in the Noor Pura area of Faisalabad was aware that there was not enough fall available for a conventional sewer and proposed the use of an interceptor tank system. Experience with later sewer projects showed that community representatives can understand a simple section of a proposed sewer and can use this as a tool to discuss options - for instance, the benefits and disadvantages of different longitudinal profiles. In Pakistan it is rare to see a technology successfully implemented in externally aided projects which has not been used elsewhere on a scheme implemented either by a municipality or a local community group. Nevertheless, community development professionals can veer too much to the view that the community always has all the answers, thus implying that there is no place for specialist engineers at the local level. In practice, there is a need for dialogue with each side bringing its own perspective and insights to bear on sanitation issues.

Conflicts between the understanding and priorities of community groups and government officials

Not surprisingly, community members normally place a high priority on constructing facilities as cheaply as possible and this may sometimes be at the expense of their long-term maintainability. Some examples of what this means in practice have already been given—for example, in relation to the need for frames on manhole covers and benching within chambers. There are also likely to be differences of opinion on permissible minimum slopes, with community members perhaps wishing to lay sewers at flatter slopes than professionals might think desirable. (Although, as we have seen, there is considerable debate among professionals on this matter.) Potentially more difficult are those differences of opinion on what constitute acceptable standards for sewer pipes, cover and manhole dimensions, with government engineers insisting on conventional standards which make sewers unaffordable to community groups.

It is still too early to be categoric about the way in which such differences in perspective and outlook can be resolved. It would seem that modifications and additions suggested by engineers should be acceptable to community members provided that:

- their benefits are clearly explained and, where necessary demonstrated; and
- the resulting cost increases fall within acceptable limits.

Where practices and materials used by community members provide a satisfactory product but differ from those specified by government, it will be necessary to develop revised standards and procedures which take into account the location and function of the facilities. In general, it seems reasonable that government authorities should accept that communities should have a major say in what is acceptable for facilities that they provide and maintain themselves, provided always that the integrity of downstream facilities is ensured.

Arrangements for project finance

The basic concept underlying the FAUP is that community members and government should share the responsibility for provision of services at the local level. In the case of infrastructure, this means that the community members will contribute 50 per cent of the cost of any scheme implemented under the project. This contribution may include some non-monetary inputs, for instance the provision of labour and materials, but substantial cash inputs have been made to all the infrastructure projects implemented to date. Once it has been agreed that a project is worthwhile and the community group has raised its contribution, a joint account is opened in a local bank with one community member and one social organizer as signatories. This account is then drawn on as required to finance the work as it proceeds.

Contract arrangements

For all the local sewers constructed to date, the community group has purchased the required materials and has hired a local jobbing contractor to carry out the work . For the first project, a 60 m length of sewer serving nine houses which cost about Rs6000, the contractor was paid on a daily basis so that the responsibility for project management remained exclusively with the community. On subsequent projects, which were rather larger, typically costing up to Rs40 000 and serving up to about 40 houses, an overall price was agreed with the contractor, based on the length of sewer and number of chambers to be provided. These arrangements are very different from government procedures and there will be a need to formalize them in the future.

It may be advisable in future to develop contracts based on simple schedules of rates for measured items, for instance the length of sewer and the number of chambers. This will allow for some flexibility if there is a need to change some aspect of the work in the course of construction—for example, the provision of an extra chamber.

Overall supervision of the projects implemented to date has been by sub-engineers employed by the FAUP, together with nominated members from the community. As the project develops, it is planned to provide opportunities for small contractors and community members to develop their skills so that eventually they are competent to undertake small infrastructure projects with a minimum of external supervision.

4.5 LOCAL SEWERAGE IN CONTEXT

The main concern of the chapter up to this point has been with the provision of sewers at the local or tertiary level. However, it is important to recognize that the provision of local sewers alone is an inadequate response to deficiencies in sanitation and drainage. At the local level, it is necessary to ensure that connections are made to the sewers, but insufficient attention to the provision of trunk and collector sewers and treatment facilities will result in problems being transferred rather than eliminated. Thus, for example, while the provision of local sewers in Orangi has transformed conditions at the plot and lane levels, it has created pollution in the nalas running through the settlement and the creeks, such as the Lyari River, to which these discharge. Local sewerage cannot therefore be treated in isolation if severe environmental problems are to be avoided. Lahore provides an example of what this means in practice. In a city of over 4 million people, there is at present no sewage treatment and most of the city's sewage finds its way to the River Ravi, resulting in severe oxygen depletion downstream. While it is undoubtedly true that the hazards produced by this situation are significantly less than those which would ensue if no action had been taken to improve sanitation and remove sewage from residential areas, it is far from ideal.

The cost of secondary sewers in North-east Lahore, excluding the main trunk sewers and treatment facilities was about Rs1700 per household, bringing the total cost for secondary and tertiary sewers plus house connections to at least Rs3000 per household. The cost of trunk sewers and sewage treatment will add to this figure, making the overall cost of sewerage greater than that of on-plot sanitation, which in Pakistan is about Rs3000 per household. However, most on-plot sanitation systems do not cater for sullage water and the cost of providing sullage drains must be taken into account when evaluating on-site sanitation options. The point to take from this is that sewerage options should not be appraised on the cost of local sewers alone.

Regardless of the above, the lack of off-site sewerage is often an absolute constraint on the introduction of conventional sewerage at the local level. Problems arose in North-east Lahore because there was no trunk sewer available to take flows from the collector sewers constructed in the course of the project. Although such a sewer was under construction, its completion was delayed when the contractor encountered technical problems. In the meantime, a temporary pumping station was constructed under the North-east Lahore project but this only relieved the problems of part of the area. There were also some difficulties in getting the Water and Sanitation Agency to provide the resources to operate the temporary pumping station and this illustrates the problems which are inherent in projects which are made the responsibility of specially formed 'upgrading units' rather than the appropriate line agency. In Faisalabad, three of the four pilot areas are served by collector sewers but these all have limited capacity and discharge to pumping stations from which they are lifted into open channels without any treatment. These situations illustrate the point that off-site facilities must be fully considered when sewers are provided in an area. Where such facilities are not available, as will often be the case, other options such as sewered interceptor tank systems which may be connected to existing drains will have to be considered.

In cities such as Lahore, Faisalabad and Madras, the topography is such that sewage discharged into conventional systems has to be pumped at least once in the course of disposal. Pumping uses energy and therefore requires financial resources which are usually in short supply. Sewage pumping stations in Lahore and Faisalabad do not operate 24 hours a day. In Lahore, pumping is often discontinued overnight and in both cities there are frequent periods of electrical 'load shedding' during which there is no power for the pumps. Many of the pumping stations in Faisalabad are poorly laid out and are in such a very poor state of repair that they must break down frequently. Observation suggests that operators frequently run pumping stations with the wet-well liquid level above the soffit of the incoming sewer as a matter of routine. The result is that the incoming sewers are surcharged, so that design assumptions on flow velocities are not realized. It would be surprising if this pattern of operation were not to be found throughout the subcontinent. One result of this is that settlement is likely to occur in main sewers, reducing their capacity in the short term and their working life in the longer term. This adds to the problems which most rapidly developing cities face with the lack of capacity and the limited extent of existing systems.

There is often a large gap between the long-term objectives of ambitious sewerage master plans and the immediate needs of local communities. For instance, none of the four pilot areas in Faisalabad will be affected by current investment in sewerage and sewage treatment until further investments are made in trunk sewerage. Provision for these investments is included in plans for the next century, but experience to date suggests that financial constraints are likely to lead to considerable delays in the implementation programme. There are no easy solutions to this situation. In theory, there is an argument for decentralizing systems as much as possible so as to reduce the length of trunk sewers. At one extreme, the Yusufabad and Thirukatchur schemes are completely independent of centralized systems. In Yusufabad, some treatment is provided in a series of septic tanks, each receiving the effluent from 60 to 80 houses but a better approach might have been to buy sufficient land for a waste stabilization pond system. Provision for local treatment is provided in several of the TNUDP sites-and-services schemes.

There are two potential problems with this approach, the first that of obtaining the land necessary for treatment and the second of ensuring that local treatment facilities are operated and maintained. The first is a real issue in Pakistan because of the high price of land around large cities. Land prices around Yusufabad are currently about Rs300 (US$10) per m^2, and prices around cities like Lahore are higher. Both governments and community groups are likely to be reluctant to spend scarce resources on land for treatment, given that the provision of sewerage solves the local environmental problems which most concern people. At Yusufabad, the possibility of combining waste stabilization pond treatment of the effluent from septic tanks with fish farming was considered and a scheme was prepared but it was eventually decided that the residents of the settlement were insufficiently well motivated to undertake the management of the fish farming scheme.

The discussion above assumes that waste stabilization ponds are the appropriate treatment technology for disposal of sewage at the local level. There is considerable evidence that the operation of conventional treatment works is often less than satisfactory because of poor operation and maintenance and it would seem extremely doubtful whether conventional technology can be introduced at the local level. However, while inadequate operation and maintenance often contribute to poor performance, there is another factor which is often overlooked and which potentially applies to waste stabilization systems as well as conventional treatment works. This is the tendency for facilities to become overloaded because investment levels are insufficient to deal with rapid development. An example is provided by the Hayatabad treatment works in Peshawar which serves a large and generally successful government sites-and-services scheme aimed primarily at middle and upper income people. The works consists of aerated lagoons followed by facultative and maturation waste stabilization ponds. Performance was satisfactory for about 3 years and then began to deteriorate as flows from areas which had not been allowed for in the original design were diverted to the works. Land had been set aside to extend the works but there was no political will to provide finance for the extension. The deterioration in performance showed itself mainly in the form of increased odour problems and after complaints from local residents, much of the flow was diverted around the works without treatment, thus removing the reason for the works being there. (It is worth noting, however, that the facultative ponds appeared to operate satisfactorily at rather higher loadings than is assumed by current process design guidelines.)

4.6 CONCLUSIONS

The first broad conclusion to be drawn from this chapter is that approaches and technologies exist to provide sewerage at the local level, making it an affordable option for low-income communities in many circumstances. More needs to be done to convince governments that appropriate standards are technically acceptable. Community involvement in sewer planning and construction is possible at the local level but government or some other agency must provide some technical and organizational support, at least in the early stages of any programme. In general, government will have to retain the responsibility for the provision of higher level facilities, but the ability to extend these facilities to keep pace with rapid development is often limited. There are theoretical reasons for decentralizing sewage disposal systems to a considerably greater extent than would be sensible in high-wage economies, but there are many practical questions to be resolved about such decentralization. This suggests that a main focus of research over the coming years should be on obtaining answers to these questions.

4.7 REFERENCES

Balfours and Engineering Consultants (1987). *Feasibility Study for Preparation of Sewerage and Waste Water Disposal Project in Karachi—Draft Final Report: Volume V—Low Cost Sanitation.* Karachi: Karachi Water and Sewerage Board.

British Standards Institution (1985). *British Standard Code of Practice for Building Drainage (BS8301: 1987).* London: BSI.

Building Research Establishment (1984). *Access to Domestic Underground Drainage Systems.* Technical Digest No. 292. Watford: Building Research Establishment.

Cotton, A.C. and Franceys, R. (1987). Sanitation for rural housing in Sri Lanka. *Waterlines,* **5**(3), 9–11.

GHK and WEDC (1991). *Evaluation of Upgrading Initiatives in Pakistan.* Final Report for ODA. Loughborough: WEDC, University of Technology.

Lillywhite, M.S.T and Webster, C.J.D. (1979). Investigations of drain blockages and their implication for design. *The Public Health Engineer,* **7**(2), 53–60.

Mustapha, S. (1985). Low cost sanitation in a squatter town: mobilising people. *Waterlines,* **4**(1), 2–4.

Naylor Bros (Clayware) Ltd (1982). *Building Drainage Design Manual.* Barnsley: Naylor Bros.

Otis, R.J. and Mara, D.D (1985). *The Design of Small Bore Sewer Systems.* TAG Technical Note No.14. Washington, DC: The World Bank.

Pickford, J.A. (1983). Planning for improved sanitation. In *Between Basti Dwellers and Bureaucrats—Lessons in Squatter Settlement Upgrading in Pakistan* (ed. Schoorl, J.W., van der Linden, J.J. and Yap, K.S.), pp. 205–218. Oxford: Pergamon Press.

Pickford, J.A. (1990). Community infrastructure in some Asian cities. In *Planning, Shelter and Services for the Urban Poor (Proceedings of the 7th Inter-school Conference on Development),* pp. 61–66. Loughborough: WEDC, University of Technology.

Tayler, W.K. (1990). Sewerage for low income communities in Pakistan. *Waterlines,* **9**(1), 21–23.

Tayler, W.K. and Cotton, A.C. (1993). *Urban Upgrading: Options and Procedures for Pakistan.* Loughborough: WEDC, University of Technology.

UNCHS (1986). *The Design of Shallow Sewer Systems.* Nairobi: United Nations Centre for Human Settlements.

Water Services Association (1995). *Sewers for Adoption: A Design and Construction Guide for Developers,* 4th edn. Medmenham: Water Research Centre.

World Bank (1994). *World Development Report 1994: Infrastructure for Development.* New York: Oxford University Press.

Yap, K.S. (1993). *The Urban Poor as Agents of their own Development: Community Action Planning in Sri Lanka.* Nairobi: United Nations Centre for Human Settlements.

5

Operation of Sewer Systems in Ghana

David M. Brown

5.1 INTRODUCTION

During a recent short visit to Ghana, a review was made of the operation of sewerage systems and recommendations were produced for how a new simplified system should be operated by a potential private contractor to ensure a high standard of service for its users. This chapter examines the current operation of sewer systems in Ghana, and the resultant standards of service they deliver to the users. It discusses why this fails to deliver an acceptable standard of service and then proposes a different operational strategy, based on a programme of planned maintenance supplemented inevitably by reactive maintenance.

Ghana is situated in sub-Saharan Africa, surrounded by Côte d'Ivoire, Burkina Faso and Benin. It gained independence from the UK in 1957, and continues exporting cocoa, timber and precious minerals. It has approximately 15 million inhabitants living in an area of approximately 240 000 km² with a 1988 GNP figure of US$400 per capita, and a tropical monsoon climate. The natural vegetation is thus tropical with dense rain forest tailing off into savannah and grassland in the north.

Living conditions in the vast majority of urban areas are very crowded. About 90 per cent of households live in apartment buildings with other households, and about 90 per cent of these households live in a single room. The average size of a household is about 4.6 persons, and there are approximately 50 households in each apartment building. There is often no room for people to cook, wash or bathe within the apartment room, so these activities are carried out in the courtyard of the apartment building, or along the public highway.

The large cities such as Accra, Kumasi and Tema have relatively good roads, and electricity and water supplies. However, the development of adequate safe sanitation facilities has not followed the same pattern of development as other services, and consequently the current situation is inadequate.

5.2 CURRENT OPERATIONAL PRACTICES IN GHANA

A small survey was carried out on the sewerage system in Tema, a port city built about 30 years ago. It is under a poor state of repair, mainly due to a lack of programmed maintenance. Although it is approximately 30 years old, the problems are typical of a much older system. As Tema is a relatively new city, the Tema Development Corporation (TDC) was responsible for the sewerage system, although Tema Metropolitan Assembly (TMA) is expected to assume these duties in the near future.

In practice about 95 per cent of the sewers suffer from sand deposition. Approximately 40 per cent of these pipes suffer a loss of cross-sectional area greater than 50 per cent. Thus at least half of the original flow capacity has been lost in these pipes, and consequently the so-called 'self-cleansing velocity' or 'minimum tractive tension' is not achieved and further deposition occurs. In addition to loss of capacity, the problem is further exacerbated by the addition of bulky materials, via the WC, into the sewer. Frequently the disposal of such materials leads to blockage formation.

Blockages can take between 8 and 20 weeks to clear from the moment they are first identified. The loss of WC facilities or the emergence of sewage either into the surface water drain or from the inspection chamber is often the moment it is recorded. The householders then contact the direct labour department of TDC and the report is logged. Due to a large backlog of work, reports are not immediately attended to, although priority seemed to be given to persistent complaints.

Three distinct catchments drain to pumping stations, which have emergency overflows from their wetwells into local water courses. Two of the pumping stations work sporadically due to pump maintenance difficulties, and there is evidence of frequent pollution incidents due to sewage flowing through the emergency overflows into local water courses. This is not perceived as a problem by the operators of the pumping stations, and the municipal engineer appears not to have highlighted it as such. As the third pumping station, which mainly serves the industrial catchment, does not operate due to theft and vandalism, the flow is discharged directly into a local water course, where approximately 75 per cent of the combined flow is raw sewage. This water course passes through agricultural and residential areas and so poses a severe threat to local public health. As this sewage may also contain heavy metals and toxic chemicals, it is unquantifiable what long-term effects this may have on the population, either directly or via the groundwater, which serves as a source of potable supply.

The rising mains that emerge from the three pumping stations are designed to pass for about 1 km under pressure, before the effluent flows under gravity, via a grit collection chamber, to a short sea outfall. However, the three pipes are badly corroded in parts, and sewage escapes into the environment. There has been no money in the past for the rehabilitation of these mains, leading to rapid deterioration. Often small-scale crops are grown alongside these water courses, with irrigation water provided directly from them. It is a cause for concern that the majority of these crops, such as lettuce and other salad crops, will carry pathogens directly to the consumers of these products.

A grit collection chamber, which had become disused over the last 10 years, because of the problems with the pumping stations and the rising main previously discussed, was rehabilitated in early 1995, when approximately 200 m³ of solids were removed. It is an open chamber, constructed from mass concrete to a Soviet design, and is approximately 20 m wide by 35 m long, and about 3 m from ground level to invert level. It contains a dry weather flow channel, and a crude concrete flow attenuation device at the downstream end. It is difficult to conceive that this channel could ever operate effectively, due to its inherent bad design, and given the lack of operation and maintenance culture within the municipal assembly and the local development corporation.

Flow should pass through the grit channel to be discharged at sea via a short sea outfall. However, this outfall is now fractured in a number of places, resulting in premature disposal close to the shore and hence inadequate dispersion. The fractures were caused by the practice of illegal wrecking of unwanted ships along the shoreline in this area of the coast. Approximately 20 wrecked small ships and boats were counted along a 3 km stretch of coast to the east of Tema.

The maintenance of the sewerage system in Tema is operated by the TDC, which uses a labour gang to perform local dig-down repairs and rodding operations. They use pitch fibre pipes to replace structurally inadequate or completely blocked sewers which do not respond to rodding. Rodding is initially carried out on blocked sewers, and experience suggests that a combination of sand and inappropriate materials discharged down the WC cause the great majority of blockages. The steel rods that are used readily clear blockages that contain items which a corkscrew-type tool can cope with. Thus the rods are used in both tension and torsion as the tool rotates. Sand deposits are difficult to displace with this type of approach.

It is clear from an examination of these expensive First World technologies that they are inherently unsuitable for use in developing countries. However, in areas of high-density housing where there are individual house connections to the water distribution network, some kind of sewerage is not only appropriate, but essential.

5.3 THE ASAFO SIMPLIFIED SEWERAGE SCHEME

Asafo is a central district of Kumasi with an estimated population of 20 000 people that has recently been sewered using simplified sewerage. This is part of the Kumasi Sanitation Programme which aims to completely eradicate the deplorable practice of public pan latrines. The main lines of the Asafo simplified sewerage system were laid out among the buildings roughly perpendicular to the local contours to take maximum advantage of the available gradients. The Asafo schemes was designed with the features described in Table 5.1.

With inspection chambers costing less than manholes due to reduced excavation, materials and construction costs, the savings made compared with a conventional sewerage scheme are in the order of between 20 per cent and 50 per cent (Bakalian *et al.*, 1994). Conventional sewerage design requires a minimum level of cover that requires extensive excavation. The Asafo project had a minimum level of cover of 0.5 m in non-trafficked areas, and

Table 5.1 Design features of the Asafo simplified sewerage scheme

Design feature	Conventional sewerage	Asafo simplified sewerage
Change of slope	Manhole	Inspection box
Change of diameter	Manhole	Inspection box
Pipe junction	Manhole	Inspection box
Connection	Upstream inspection chamber	Upstream rodding eye
45°–90° bend	Manhole	Inspection box

1.0 m below trafficked areas. As none of the sewer lines passed under heavily trafficked streets, this depth of cover was sufficient in conjunction with PVC pipes. PVC pipes offer the advantage of longer lengths, and so fewer joints and thus less opportunity for poorly made joints to initiate blockages. Plastic gasket joints were used with the PVC pipes.

Economic circumstances has prevented the great majority of households from purchasing an appropriate WC for connection to the sewerage system. A community education campaign has been recently carried out which emphasizes the benefits to public health, the householders and the nightsoil labourers that may be achieved once the transition is made from pan latrines to the simplified sewerage system.

5.4 PROPOSED OPERATIONAL PRACTICES

A new manual of operational and maintenance practices was developed for the new simplified sewerage scheme. This was widely based on the Water and Sanitation Agency (WSA)/Foundation for Water Research (FWR) (1991) publication *A Guide to Sewerage Operational Practices*, and due acknowledgement is made to WSA and FWR, although it was adapted specifically for Ghana, and also for developing countries in general. This section discusses some of the most important issues which the manual addresses.

5.4.1 Ethos of proposed operational practices

In order to develop an effective sewerage operational policy for a sewerage system it is necessary to have a clear understanding of

the objectives involved. For sewerage operations and maintenance in Ghana, typical aims are:

- to comply with safety and public health requirements;

- to ensure that the best operational knowledge and experience
  is utilized whenever practical in the design of new works;

- to ensure that all work is carried out in the most cost-effective
  manner;

- to ensure the integrity of each element of the sewerage system;

- to minimize nuisance from odours and noise;

- to ensure that failure of any part of the sewer scheme has the
  minimum effect on the rest of the system; and

- to ensure that only authorized persons enter or work on the
  sewer system.

5.4.2 Plannng a maintenance strategy

The size of a typical simplified sewerage system makes it infeasible and uneconomic to give frequent attention to all sewer
lines. In spite of this general lack of maintenance, many simplified sewers continue to function adequately and do not give rise
to problems. Because of this, there is great scope for introducing
a system of inspection and maintenance that identifies only those
parts of the system which justify attention on a predetermined
programmed basis, leaving the remainder to be dealt with only
on a reactive basis. Often in developing countries, only a small
minority of the sewerage system should be dealt with on a
planned basis.

Reactive or crisis maintenance involves responding to failures
and problems as they occur. In the less sensitive parts of the
simplified sewerage system, where failures or problems do not
have severe consequences, reactive maintenance is the most
appropriate system because it will generally be cost effective. In
contrast, planned maintenance involves undertaking inspections,
and carrying out maintenance in order to reduce frequency or
risk of failure. If it does not achieve these objectives, then the
additional expense of planned maintenance cannot be justified. It
must give 'value for money' to the sewerage operator.

Planned maintenance should not be confused with routine maintenance. Planned maintenance involves establishing which sewers, or other assets, should be cleaned and the optimum sewer cleansing frequency. This requires good management, efficient data storage systems and regular performance monitoring with a view to improving the system of planned maintenance so that the right balance is achieved between the levels of service achieved and their costs.

5.4.3 Data requirements

The manager of a sewerage system requires data on sewer blockages, rising main failures, pumping station failures, actual response times, flooding incidents and loss of WC use, as well as planned maintenance records for effective control. The data should be stored in such a way that they can be quickly retrieved and analysed appropriately. Although there are many suitable computer database packages to do this on a personal computer, it can be quite adequately done with an indexed manual data storage and retrieval system.

5.4.4 Term contracts

Term contracts offer the client, who has what is often a variable and unpredictable workload, much greater flexibility in terms of manpower and plant resources. The contractor may be the direct labour force of the local authority issuing the contract or, as is increasingly happening in Ghana, a private contractor. Contracts should be for a minimum period of 12 months. Contracts for longer periods of up to 5 years can save the administrative expense of re-advertising, re-tendering and assessing the tender. With longer contract periods, however, the effects of inflation will usually need to be incorporated. This may be by predetermined price increases in subsequent years, or by inflating prices in line with the inflation figure. The Institution of Civil Engineers (1988) *Conditions of Contract for Minor Works* is particularly appropriate in Ghana, as local contractors and civil engineers are very familiar with it.

Typically, a schedule of rates for general civil engineering sewer maintenance work, will include sections covering daywork rates for labour, general items, earthworks, *in situ* concrete, concrete ancillaries, pipework, brickwork, blockwork and masonry; painting; sewer renovation and ancillary works. Contracts are usually awarded to the contractor who offers the lowest rates. A tender total can be calculated in the normal manner if weighting factors are included in place of quantities, indicating the likely ratio of the work to be undertaken associated with individual items.

5.4.5 Technical training

Good management decisions cannot be made without a sound technical understanding of the sewer network and how it performs under different conditions. Training programmes should include training in such areas as pumping station behaviour, and repair and rehabilitation techniques for sewers. Manual staff are able to carry out their duties more effectively if they are given basic instruction in the operation of the sewerage system and its ancillaries, and this is especially true where the sewerage system has recently been introduced, such as in Asafo.

5.4.6 Health and safety

There is a moral responsibility for employers to ensure the safety of their employees, notwithstanding national laws and regulations. There may also be laws governing the use of substances that may be hazardous to health which employees should be made aware of. It is therefore necessary to establish formal procedures which, if followed, will ensure that the legal requirements are met, namely:

- that the site of the maintenance work is safe before operatives enter on to or into it, and remains safe while the work is carried out;

- that third parties, including contractors and members of the public, are warned of any potential dangers, and are adequately protected from them; and

- that the work is carried out safely by suitably trained and supervised operatives following accepted safe working practices.

It is extremely important that where there are mains water supplies to sewerage installations, they should only be connected by pumping from a storage tank. Local laws should reinforce this action to prevent any possible contamination to the public water supply.

5.4.7 Operational problems

Recommended procedures

Sewer cleaning and blockage clearance work will typically involve a mixture of reactive and planned maintenance work. Reactive maintenance will be needed to clear blockages of debris which may be causing localized flooding or restrictions on toilet use. Where there are persistent problems, it may be appropriate to carry out sewer cleaning of some lengths of pipe on a planned maintenance basis. It may also be necessary to clean out some flow control devices such as those on storm overflows, grit collection chambers and detention tanks after operation of the installations. The following procedures are recommended for sewer cleaning work.

1. Steps should be taken to prevent sediment and debris from entering the sewer system. This can be achieved by ensuring that builders take appropriate action during the construction works. Careful thought should also be given to the design of inspection chambers and manholes, to prevent the possibility of sand and dust entering the system through these appurtenances.

2. Where sediment problems have occurred and recurrence is considered a possibility, periodic inspections should be carried out to determine current sediment depths and the rate of deposition. Where the problem recurs, planned maintenance should be considered.

3. The sewerage manager will need to have procedures for

dealing with blockage reports and mobilizing resources to deal with them within target response times.

4. In order to develop and maintain cost-effective planned maintenance programmes, appropriate methods should be maintained of all sewer cleaning work. Records of sediment depths from sewer, manhole and storm sewage overflow inspections should also be kept and used for this purpose.

5. In cases where there are persistent problems, the sewerage manager will also need to assess whether it is cost-effective for him/her to continue to clean a defective sewer, or whether to invest in improvements. In arriving at this decision, the sewerage manager will have to balance economics against the levels of service failure.

It is recommended that surveys are carried out to determine the effectiveness of sewer cleaning work. However, such surveys need not include full recording of defects.

Sedimentation

The term 'sediment' covers a wide variety of materials from fine solids, carried initially in suspension at some stage may settle in the sewer invert, to the large 'obstacles', which are either stationary or roll along the sewer invert. Debris is an additional associated problem. This may include bricks, rubble, rags and other sanitary refuse. Sediment can find its way into the sewer system in a number of different ways: through toilets, kitchen connections, inadequate pipe joints, manholes and inspection chambers. Construction sites can prove particularly troublesome sources of debris. Sediments will generally be cleared by jetting, but occasionally rodding, winching or hand excavation may be necessary.

Tree roots

Any sewer, particularly where there are defects such as open joints or fractures, may be affected by tree roots attracted by the source of moisture. Roots may be classified as fine, mass or tap roots. Mass roots cause the greatest problem, significantly reducing flow capacity, and are the major contribution towards sewer blockages. However, tap roots may be the most difficult to remove. A local excavation is necessary for removing root intrusions. In order to minimize the effects of roots on the sewer, trees should be planted at least 6 m away from the line of sewers.

Blockages

A blockage is a full or partial restriction within the sewer, which may result in flooding or premature operation of overflows. Blockages can occur on any sewer length, but are more frequent in small diameter sewers. Most blockages require a specific maintenance action, e.g. rodding to clear them, but on occasions the flushing action caused by multiple WC operations may move the blockage further down the system to a location where it becomes less significant.

Intruding connections

Poor constructional practice in Ghana has resulted in numerous instances of new connections being made to existing sewers where the lateral connection protrudes through the pipe wall and into the sewer. This reduces cross-sectional area, and may lead to blockages. In some instances the structural damage caused to the sewer, by a poor connection, may lead to deformation and eventual collapse. Intrusions can also be caused by the careless laying of services through sewers. This also reduces cross-sectional area, and can cause safety problems. Any reduction in cross-sectional area will restrict the flow, and under certain conditions, can result in a build-up of debris which must be removed. Excavation is the most appropriate form of remedial action to remove intruding connections.

Grease

Grease is generally associated with non-domestic users, e.g. hospitals, schools, restaurants, etc., and is often due to inadequately sized, badly designed, inadequately maintained or missing grease traps. A high-temperature effluent containing grease will usually cause problems as the effluent starts to cool. The grease will then adhere to the wall of the sewer, causing a reduction in capacity. Grease may be removed by jetting, rodding, winching or hand excavation. Grease removal from sewer systems and pumping stations is an expensive operation and it is often more cost-effective to identify the source and take action to prevent further discharge of greasy effluents.

Encrustation and scale

This is the build-up of material typically found on the walls of sewers. The consistency and thickness of the material build up on the sewer wall can vary significantly, either as a result of chemical reaction between the sewer material and the sewage, or due to infiltration into the sewer. The encrustation should only be removed if it is causing a problem, as removal always carries a risk of damage to the fabric of the sewer.

Sewer collapse

Full or partial collapse will obviously result in problems. The structural integrity of the sewer can be severely weakened by the loss of part of the pipe fabric, which may collapse into the sewer, and result in a blockage. This situation may be made worse if repairs are carried out to the sewer, and the defective pipe, surplus construction material or fill is not removed from the sewer, inspection chamber or manhole.

Infiltration

Defects in the sewer fabric or jointing can lead to the ingress or egress of water and solids. Fines, or other solids which may be carried in by the infiltration water, can settle out and reduce the hydraulic capacity of the sewer or cause blockages.

Trade effluent

Under the specific by-laws of individual countries, discharges of 'trade effluent' should be controlled so as to reduce their impact on the sewer system, the sewage treatment facility and therefore the environment. Therefore, trade effluent should not normally cause cleaning problems other than when a pre-treatment plant fails and excessive solids (e.g. slaughterhouse blood and offal) are passed into the sewer system.

5.4.8 Methods for cleaning sewers

Rodding

Rodding is a manual technique for the clearing of blockages that occur within small-diameter, shallow sewer systems, generally not exceeding 225–300 mm diameter and less than 2 m deep. For sewers greater than 300 mm diameter, the rods tend to wander within the sewer, and are not very effective. The distance from the access point (manhole or inspection chamber) is limited to about 25 m, above which the operation becomes ineffective.

Rods are generally of 1–2m lengths of steel, plastic or cane which are screwed together to form a semirigid construction. The screwed joint should not be capable of becoming detached during operation. Although the steel and cane rods operate much more effectively in tension and torsion, plastic rods are extremely effective when working in compression, in order to push the blockage material to the next manhole. Consequently the most effective strategy is the complementary use of both types of rods, with the choice dependent on the type of blockage. This decision may be readily made by an experienced operator.

Personal experience in Ghana suggests that rodding is probably the most cost-effective method of alleviating blockages within specific parts of the sewer system. If a blockage cannot be cleared by rodding, then jetting or another suitable technique should be used. The benefit of rodding is that it requires minimum labour and capital costs. However, it is essential that the operators should remove, whenever possible, the debris causing the blockage.

Jetting

Sewer jetting involves the eroding and removal of sediment or blockage material from the sewer system. Jetting machines can clear sewer blockages, and erode sewer sediment within the pipe to a point where removal can take place.

Winching

Winching or dragging is the technique of pulling a shaped bucket through the sewer pipe. The bucket collects sediment which is then emptied out at the point of access, normally into a skip.

The winch operation can be done by hand, but in most cases the winches are power driven. Winching is most effective in heavily silted sewers up to 900 mm diameter. Winching is used infrequently because it is very slow, and it can often cause damage to the fabric of the sewer.

Hand excavation

Traditionally, the larger diameter outfall sewers were cleaned by manual methods, i.e. by labourers shovelling sediment into skips which were transported through the sewer to the manhole and thence removed. However, as the sewer environment is hazardous, the number of occasions that operatives are required to work in sewers should be minimized.

Solids removal and disposal

It is important that the displaced solids or sediment are removed from the sewer system, and disposed of at a licensed tip.

Local pipe replacement

Depending on local labour costs, local excavation and pipe replacement may be cheaper than some of the above methods, and should be considered. Care should be taken to present additional debris entering the sewer system during this work.

5.4.9 Rodent control in sewers

Rodent control in sewers is not only needed to minimize health risks, but to prevent structural damage to the sewers, inspection chambers and manholes, resulting from the burrowing of rodents. The objectives of rodent control in sewers are to institute control programmes to minimize rodent damage to the fabric of the sewer system, damage to the environment, and any risk to health and safety.

5.4.10 Sewer tracing procedures

The technique of adding a dye to the flow in order to be able to identify it downstream is well known. The method is primarily used for determining the route of a sewer by adding dye to the upstream flow and observing where it appears downstream. Whenever flows are low it may be necessary to add water from a tanker or hose to temporarily increase the flow. An alternative method, which is commonly used in Ghana, where dye is not readily available, is to use noise. One operative will gently but repeatedly tap the inside of an inspection chamber while a colleague will listen at another; with experience an operative can readily distinguish a directly connected chamber from an unconnected one.

5.4.11 Inspection chambers and manholes

Inspection chambers and manholes are constructed in sewers to provide access for maintenance and inspection of the sewer. They are generally constructed in sandcrete blocks or *in-situ* concrete, and are sometimes provided with step irons or ladders for access to the sewer. The sewerage manager should ensure that:

- manholes and inspection chambers are structurally sound;
- authorised persons can enter manholes in a safe manner, and in compliance with the sewerage undertaker's safety procedures; and
- manhole and inspection chamber covers are not a hazard.

A reactive maintenance strategy will generally be most appropriate for the maintenance of manholes and inspection chambers. Maintenance work will therefore only be carried out when a problem is reported or otherwise discovered. The routine replacement of manhole covers and frames because of age, expected asset life, etc. to a routine programme, is unlikely to receive sufficiently high priority, and is not considered cost-effective.

Inspection following complaints or in advance of planned access for maintenance works may be appropriate in some circumstances. The general inspection of manholes or inspection chambers for the sole purpose of determining their condition is not cost-effective and is not therefore recommended. Whenever a manhole or inspection chamber is entered for any purpose (for example, to clear a blockage or for inspection of the sewer), an inspection should be carried out and a report completed.

It is possible that the need for replacement of covers and frames will be identified from reports from the public highway authorities or from observations from employees involved with day-to-day maintenance of the sewerage system. When a report is received, a list of any work required should be made and given a priority. The work should then be scheduled to be carried out within the response time appropriate to the priority assigned. All reports should be stored and monitored. Where a significant number of defects are reported in a small area it may be cost-effective to carry out a survey of all the manholes and inspection chambers in that area with a view to carrying out a planned maintenance programme.

Repairs of inspection chambers and manholes should be done to the same specification as new chambers. The methods involve work in confined spaces, and appropriate safety precautions should be strictly enforced and observed.

5.4.12 Suggested schedule for inspection of manholes

During an inspection of a manhole the following items should typically be included:

- check that the cover or frame is not broken or cracked, that it does not rock more than 10 mm, and that the eyes and keyways are not excessively worn;

- check that the cover is flush with the ground level within specified tolerances;

- check that the opening in the cover and access shaft is sufficient for safe access (it is usually considered adequate where there is a clear opening of 600 mm or more);

- check that the step irons are sound and there are no missing step irons;

- check for gas and oxygen deficiency;

- check for odour indicating hydrogen sulphide (bad eggs) and other odours;

- check the general internal structural condition of the manhole;

- check the depth of sediment in the invert;

- check that the sewer is operating satisfactory;

- record evidence of rodent infestation;

- check ladders and platforms;

- check for gross infiltration in the sewer flow; and

- check for signs of infiltration through the walls of the manhole.

This can all readily be done in the form of an operator filling in a preprinted form containing a checklist.

5.4.13 Layout of connections

A large proportion of sewer maintenance problems are caused by poorly made lateral connections. This may cause structural failure of the sewer or blockage due to a protruding lateral. A detailed discussion of this subject is given elsewhere (Bakalian *et al.*, 1994). The sewerage undertaker may allow a builder to make a connection or may elect to carry out the work himself and recharge the builder. In either case the sewerage undertaker should supervise the construction of all connections to existing sewers to ensure that it is properly carried out. Completed connections should be inspected externally prior to backfilling.

5.5 CONCLUSIONS

These recommendations were offered to the Kumasi Metropolitan Assembly, in the form of an operational manual for the Asafo Simplified Sewerage System. It is essential that the success of this operational and maintenance policy is reviewed, and the manual is updated to reflect the practices that are either not covered or incorrectly covered in the manual. It is certainly debatable whether this manual could be successfully transported to similar systems in other parts of the world, although it is expected that the principles involved will be similar.

ACKNOWLEDGEMENTS

The author wishes to thank the Institution of Civil Engineers and the Royal Academy of Engineers for their financial assistance and the Kumasi Metropolitan Assembly for its co-operation which made this work possible.

5.6 REFERENCES

Bakalian, A., Wright, A., Otis, R. and Azevedo Netto, J.M. (1994). *Simplified Sewerage: Design Guidelines*. Water and Sanitation Report No. 7. Washington, DC: The World Bank.

Institution of Civil Engineers (1988). *Conditions of Contract for Minor Works*. London: Thomas Telford Ltd.

WSA/FWR (1991). *A Guide to Sewerage Operational Practices*. Swindon: Foundation for Water Research.

6

A Low-cost Sewerage System for a Small Rural Community on a Greek Island

G.E. Alexiou, G.G. Balafutas and S.Ath. Panteliadis

6.1 INTRODUCTION

This chapter describes the first low-cost sewerage system to be designed and constructed in Greece. The aim of the project was to design an innovative system to collect and treat wastewater to enable it to be reused for irrigation. The site of the project was the settlement of Finikia, part of the Oia Community on the island of Santorini.

Thira, the local name for Santorini, is a crescent-shaped island, the remains of a land mass which was largely submerged after the eruption of the volcano which lies in the middle of the island's bay (Figure 6.1). The island is administered by the Cyclades Prefecture, which comprises 24 inhabited islands in the Aegean Sea, southeast of Athens. Thira has an area of 7500 ha. There are 19 communities on the island with a total population of 10 553. The population density is among the highest in the country, with 113 people per km^2, compared with 36.2 and 77.7 for Cyclades Prefecture and Greece, respectively.

As well as its distinctive natural landscape, the island is classified as a place of special archaeological and historical interest and as such has become an important tourist attraction. The island has 7800 tourist beds (not including camp sites), representing 21.4 per cent of the total tourist accommodation of Cyclades Prefecture. The tourist season runs from April to

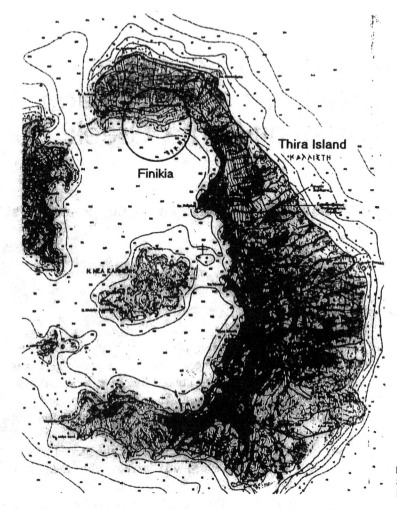

Figure 6.1 Map of Santorini showing the location of Finikia

October. The recent expansion in the island's tourist industry has been rapid, with the number of beds increasing each year by 25 per cent, on average, between 1981 and 1991. It is expected that within 20 years the current number of beds will have doubled, and within 40 years will have trebled.

6.1.1 Climate and geography

The climate of Thira is semi-arid, with more than 150 dry days each year. The mean annual temperature is 17.8°C, with a mean annual rainfall of 334 mm. Consequently, there is low crop productivity.

A total area of 3740 ha is cultivated, of which only 6 per cent is irrigated. The groundwater is of low quality, mainly because uncontrolled and illegal drilling is widespread, resulting in severe salination problems. Little water can be collected from the existing aquifer. Therefore, overall there is a shortage of fresh water. The island's soil is sandy with materials of volcanic origin (ash, etc.) and a very low clay content.

6.2 CURRENT SEWERAGE PROVISION

The intense development of Thira over the past 30 years in response to increasing tourist activity has resulted in serious environmental problems, including:

- direct discharge of untreated wastewater from communities and hotels into the sea;

- uncontrolled landfills, where solid wastes are burned; and

- salination of the existing aquifers.

Oia is in the north of the island and is classified by the Ministry of Tourism and Civilization as a 'traditional' community because of its unique location and architecture. Finikia is a settlement of the Oia Community where the majority of the permanent population of Oia (590 people) lives. The problems of the settlement can be summarized as follows:

1. *High population density*. The majority of the permanent population lives in an area of about 2.4 ha. Because the terrain slopes steeply, there are few areas suitable for building and therefore houses are crammed together very tightly, with no room for individual on-site sanitation. In some areas, stone retaining walls called '*yposkafa*' have been used to create level building plots. The density of housing is comparable with that in other parts of the world where shallow sewer systems have been constructed.

2. *Ground stability*. The existence of *yposkafa*, the volcanic origin of the soil and the narrow pathways combine to cause

problems of ground instability for construction purposes. This makes it infeasible to excavate to depths appropriate for conventional sewer systems.

3. *Water shortage*. The lack of water is a problem faced by the island as a whole. Every house has a water tank for collecting rainwater, which is used for cooking and drinking. Although the water network of Oia has expanded in recent years, there is insufficient water during the tourist season. Therefore tankers are still used to bring water from boreholes in the middle of the island by road. The proposed construction of a desalination plant for the Oia Community is expected to solve the problem.

4. *Topography*. Finikia has grown up alongside three drainage basins from south to north. The slopes have a high average gradient, ranging from 15 to 25 per cent. The three basins (here referred to as A, B and C) are dry all year round, as rainwater percolates immediately because of the soil type.

5. *Lack of final disposal*. A number of houses have septic tanks. However, because these are not emptied sufficiently frequently, they often overflow. The majority of the houses rely on infiltration pits, which remains a restricted disposal method.

These and other factors were taken into consideration to help determine an appropriate design for the low-cost sewerage system. The local Community Council was consulted, as was the community's technician. It was also important to take into account the impact on the new system of the proposed development of a number of hotel complexes.

6.3 DESIGN CONSIDERATIONS

6.3.1 Conventional design

In Greece the design of conventional sewerage systems is subject to the compulsory guidelines contained in Law 696/1974. These guidelines are considered conservative and their implementation can lead to very poor hydraulic behaviour of sewers in small villages. For example, the guidelines require a minimum pipe diameter of 200 mm, with a minimum velocity of 0.5 m/s. For a minimum proportional depth of 10 per cent, the minimum velocity must be 0.30 m/s, a condition that in practice cannot be achieved because the flows are insufficient. Similarly, trenches must be at least 2 m deep, which causes stability problems during excavation.

The guidelines are also expensive to implement. The vast majority of villages in the Greek islands, in isolated settings and with small populations and narrow pathways, cannot find the necessary funds to construct such a sewer network. In some cases this has meant that the construction of a sewer network has had to be postponed despite the fact that a wastewater treatment plant had been built.

The network was designed with the following in mind:

Population considerations

The population to be served by the network was calculated on: (a) the current permanent population, (b) the large number of existing and planned rooms in hotels and private houses; and (c) the planned construction of a desalination plant at Oia. Given that the design period of the project was 40 years, a trebling of the number of rooms was assumed. The calculated design population was 2500.

Water consumption

The mean daily water consumption during the 6-month summer period was estimated at 150 litres per caput per day (lcd). This value was broadly confirmed by measuring the mean daily consumption at a 30-room hotel complex, which was found to be 130 lcd. The maximum daily water consumption coefficient was estimated to be 1.5 (225 lcd). Assuming that 80 per cent of the water consumed becomes wastewater, the peak sewage flow was calculated at 0.0066 litres per second per caput.

Groundwater infiltration

This was not calculated because: (a) high density polyethelene pipes were to be used; (b) the network was to be very carefully constructed; and (c) there is no high water table.

Population densities

These were estimated at 15 and 50 m^2 per caput for the centre and periphery of the settlement, respectively.

Diameters

Initially, diameters were to range from 110 mm in the upper sections to 125 mm in the lower sections. These were subsequently changed to 125 mm throughout the network. The hydraulic design was based on Manning's equation. Diameter selection was based on a maximum proportional depth of 50 per cent. Maximum proportional depths were 30 per cent and 45 per cent in secondary and primary pipes, respectively.

Velocities

These ranged between 0.55 and 2.85 m/s with slopes of 2.9–25 per cent across the whole network. The maximum trench depth was 0.70 m, with a mean value of 0.35 m.

6.3.2 Low-cost sewer design

Phase A

The overall aim of the project was to collect wastewater from the three drainage basins of the settlement, treat it at a wastewater treatment plant (WTP) close to the settlement, and then use the treated wastewater to irrigate agricultural land in the north of the island [15 ha of barley (for animal feed) and 125 ha of vineyards].

The original design, submitted in March 1993, had the following characteristics:

- HDPE pipes with diameters ranging from 110 to 125 mm to be laid under all the main pathways of the settlement (HDPE pipe was chosen because it can be obtained in rolls of 100 m);

- the main pipes collecting wastewater from basins C and B to end at two pumping stations, passing their flow to the end of the main pipe from basin A; and

- the total flow to be treated at a WTP situated in a nearby field and then used for irrigation.

The total cost of phase A was estimated at almost Dr.17 millions.

Problems were encountered when trying to locate an appropriate site for the WTP. The main reason for this was the local people's unfortunate experiences with the existing sewage disposal system, and their reluctance to sanction the construction of a treatment plant close to the settlement. (Specifically, there is a conventional sewerage system collecting wastewater from a large area of Oia. The collected wastewater flows by gravity to a field in the north part of the island. After primary settling in a series of closed, poorly constructed rectangular tanks the wastewater overflows and creates an unpleasant, mosquito-infested anaerobic pond which overflows directly to the sea.)

The original design was amended by the Technical Services Department of the Cyclades Prefecture in the following respects:

- pipe material to be PVC with a minimum diameter of 125 mm. Only a small part of the network (110 m) was to be constructed in 110 mm diameter HDPE pipe, for demonstration purposes;

- inspection chambers to be a special commercial type, located every 50 m and covered with a heavy duty lid; and

- the pipes of the whole network to be protected with a 70 mm thick cement collar (this was in spite of the fact that the heaviest traffic is the donkey).

The cost increases caused by these alterations was offset by the fact that PVC pipe is cheaper than HDPE pipe.

Phase B

The second phase of the project focused on the optimal layout for the sewer network, and connecting the new system to the existing conventional sewer network.

The final pipe layout within the settlement retained only one of the pumping stations. The total flow was to be pumped via new pumping stations to the existing conventional sewerage system.

The proposed design offered two alternative solutions: the first incorporated two new pumping stations with a total sewer length of 810 m; and the second had three pumping stations and total sewer length of 700 m. Budgetary considerations determined that the second solution was chosen.

6.4 FINAL DESIGN

The low-cost sewerage network that was finally constructed has the following characteristics:

1. *House connections.* All household wastewater and sullage flows directly to the street sewer. Connection is achieved by collars attached to the pipe. The PVC connections have a diameter of 75 mm.

2. *Inspection chambers.* Inspection chambers with heavy-duty lids were constructed at maximum distances of 50 m to provide access for sewer cleaning. They have been installed at very shallow depths (average 0.35 m) with internal dimensions of 0.30 × 0.40 m.

3. *Drop connections.* At particularly steep gradients, a number of drop connections have been constructed. These comprise a vertical pipe (200 mm diameter) with a plastic inspection cover. To keep the pipes vertical, cement rings were placed at their highest and lowest points.

4. *Pumping stations.* Four pumping stations were finally constructed, each containing two submersible pumps. They are fully equipped with all the necessary facilities (floating switches, non-return valves, alarm) and were constructed below ground for aesthetic reasons.

5. *Overflow tanks.* Because power cuts are frequent in Santorini, overflow tanks were considered necessary for the proper operation of the network. These have been incorporated into the network at the four pumping stations. Whenever the electricity supply is interrupted, wastewater accumulates at each pumping station and overflows into the tanks. The tanks were

designed to be large enough to have a retention time of more than 7 h under conditions of maximum flow. A non-return valve at the base of the tank provides a connection with the pumping station.

6.4.1 Cost considerations

Compared with conventional sewerage systems, our low-cost shallow sewerage system proved to be very cheap. We were able to make an accurate comparison of costs because we were designing a conventional sewerage system at the same time as this project was underway. This was for the mainly agricultural settlement of Bryta, which is situated in Edessa Prefecture, 140 km north of Thessaloniki. The design population was 750 people. The peak flow factor was 1.8. The length of sewer was 4830 m, and PVC pipe was used with a minimum diameter of 200 mm and a maximum diameter of 315 mm.

The cost analysis of the 'internal' network of Finikia (Table 6.1) includes the construction cost of the sewers and two pumping stations. The 'external' network cost represents sewer costs and the costs of the other two pumping stations. The cost analysis is based on 1993 prices. The cost comparison between the conventional sewerage network of Bryta Community and the shallow sewerage system constructed in Finikia (Table 6.2) demonstrates the economy of the latter.

Table 6.1 Cost analysis of the Finikia low-cost sewerage system

Item	Price (Drachmas)		
	Internal	External	Total (percentage)
Earthworks	4 954 500	1 674 430	6 628 930 (23.4)
Pipes	9 961 500	3 687 695	13 649 195 (48.3)
Pumping stations	4 420 056	3 257 790	7 677 846 (27.1)
Inspection chambers	320 000	—	320 000 (1.2)
Total	19 656 056	8 619 915	28 275 971 (100)
Cost per m	9 270	12 367	10 038

Table 6.2 Cost comparison of the conventional sewerage scheme at Bryta and the low-cost sewerage system for Finikia

Item	Price (Drachmas)	
	Bryta	Finikia
Earthworks		
Excavation	96 122 689	
Backfilling	64 164 439	
Subtotal	160 287 128	6 628 930
Pipes	18 562 276	13 649 195
Inspection chambers	27 943 443	320 000
Supplementary works	3 027 460	7 677 846
Total	209 820 307	28 275 971
Cost per m	43 441	10 038

6.5   CONCLUSIONS

We have demonstrated the feasibility of constructing a shallow sewerage system suitable for a small rural community in Greece. The principal advantages of the system over conventional designs are its low costs and the short period needed for completion. When the system becomes operational in the near future, we will be able to gauge its effectiveness as a design guide for similar situations in Greece.

ACKNOWLEDGEMENTS

This project was made possible by funding from the European Community under the scheme MEDSPA 91-1/GR./012/GR./01 in collaboration with the Oia Community

6.6   BIBLIOGRAPHY

AGP Consultants (1993). *Design of the Sewerage System for Bryta Community, Edessa Prefecture*. Thessaloniki: AGP Consultants.

Balafutas, G.G. and AGP Consultants (1993). *Design of the Finikia Sanitary Sewer System: Phase A*. Thessaloniki: AGP Consultants.

Balafutas, G.G. and AGP Consultants (1994). *Design of the Force Main and Pumping Stations of Finikia Sanitary Sewer System*. Thessaloniki: AGP Consultants.

Nama Consulting Engineers (1993). *Environmental Impact Assessment Study for the Wastewater Treatment Plant and Wastewater Disposal of Fira Community*. Athens: Nama Consulting Engineers.

Sinnatamby, G.S. (1986). *The Design of Shallow Sewer Systems*. Nairobi: United Nations Centre for Human Settlements.

7

Settled Sewerage in Africa

D.D. Mara

7.1 INTRODUCTION

Ex Africa semper aliquid novi
Pliny The Elder (23–79AD)

The feasibility of settled sewerage *sensu* Mara (Chapter 2) was mentioned by Martin (1935):

> *If there is not enough fall to give a self-cleansing velocity in the main drain, it will sometimes be possible to put in a septic tank at the head of it. The effluent from a septic tank, being free from any solids capable of choking a drain, may safely be laid with a merely nominal fall.*

However, the earliest report on settled sewerage schemes actually installed did not appear until 28 years later (Vincent *et al.*, 1963). This report described sewered aqua-privies in Northern Rhodesia (now Zambia), which have been the subject of several other reports (Mara, 1976; Feachem *et al.*, 1978a, 1979; de Kruijff, 1978; National Housing Authority, 1979; Otis and Mara, 1985; Reed and Vines, 1991; Vines and Reed, 1991).

Settled sewerage (sewered aqua-privies) was also installed in 1965–68 in the resettlement town of New Bussa in what is now Kwara State, Nigeria (Waddy, 1971; Feachem *et al.*, 1978b, 1979; Otis and Mara, 1985; Reed and Vines, 1992), and more recently (1991–95) in seven townships in South Africa, mainly as sewered conservancy tanks (D.T. Human, 1995, personal communication; see also Smith, 1993). It was proposed for (but never installed in) N'Djamena, Chad (Black and Veatch International, 1975), and also for Tunis as a means of upgrading pour-flush toilets (*puits perdus*) (Mara, 1992).

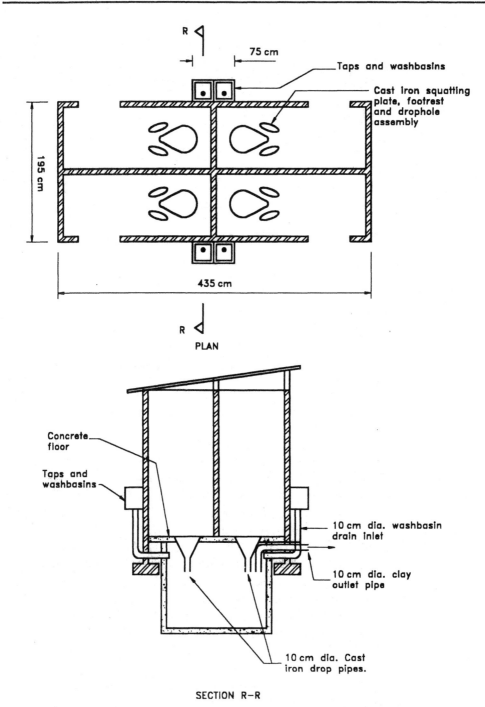

SECTION R–R

Figure 7.1   Typical sewered aqua-privy system layout in Chipanda, Matero (Zambia)

7.2 SETTLED SEWERAGE IN ZAMBIA

Settled sewerage was originally developed in the late 1950s by Mr L.J. Vincent, Manager of the then African Housing Board of Northern Rhodesia (now the Zambian National Housing Authority) (G.v.R. Marais, 1995, personal communication). Conventional and sullage aqua-privies did not work well in Northern Rhodesia and so settled sewers were developed to remove the settled wastewater (toilet wastes and sullage) from the aqua-privy tanks. The first such system was installed in 1960 at Kafue, an industrial township 50 km south of Lusaka; the land here, known as the Kafue Flats, is very flat indeed with a fall of only 1 in 2000. The sewers were designed for daily peak velocity of 0.3 m/s, and the pipes were 100 mm minimum diameter laid at a minimum gradient of 1 in 200. They were designed only to flow partially full and not, unlike the more recent North American

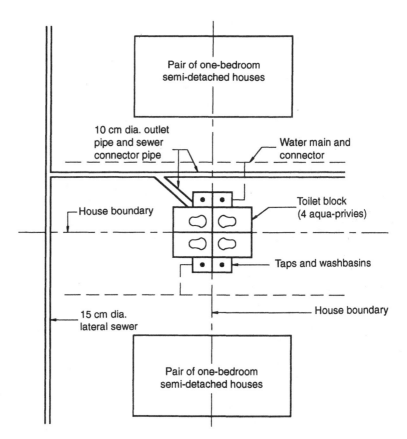

Figure 7.2 Typical sewered aqua-privy block in Chipanda, Matero (Zambia)

systems described by Otis (Chapter 9), for surcharged flow. One
Zambian system—that at Chipanda in Matero Township,
Lusaka—is described below, but several others exist and are
described elsewhere (Feachem *et al.*, 1978*a*, 1979; de Kruijf,
1978).

The Chipanda system, which was installed in 1960, serves 532
households. Each aqua-privy block serves four households, and
immediately outside its toilet compartment each household has a
water tap and a sink which discharges its sullage into the aqua-
privy tank (Figures 7.1 and 7.2). The tank effluent discharges, via
a 100 mm asbestos cement connector pipe, into a 150 mm
asbestos cement lateral sewer which runs between most of the
compounds. Originally the settled sewage was treated in a series
of waste stabilization ponds, but these were abandoned when the
settled sewers were connected to the city's expanded conventional
sewerage system.

Figure 7.3 Sanitation block in family compound in New Bussa, Kwara State (Nigeria)

7.3 SETTLED SEWERAGE IN NIGERIA

The New Bussa sewered aqua-privy scheme was installed in 1965–68: it serves 256 enclosed family compounds, each housing 15–40 people. Each compound has a sanitation block (Figure 7.3) comprising a laundry, shower compartment and an aqua-privy, with a water tap in the laundry area. Sullage is discharged into the aqua-privy tank, which in turn discharges into a short length of 100 mm diameter asbestos cement pipe which is connected, via a street junction box, to a 100 mm or 150 mm diameter collector sewer which runs in the lane or street outside the compound (Figure 7.4). The wastewater is treated in one of two single facultative waste stabilization ponds serving the east and west sides of the town.

No information on the hydraulic design of the settled sewers is available, but it can be reasonably assumed that it was based on

Figure 7.4 External view of New Bussa sanitation block showing settled sewer leading to street junction box and collector sewer

the Zambian schemes described above, as these were the only other schemes in existence at that time. User satisfaction was generally high, the main complaints being the intermittence of the water supply and the need for manual desludging of the aqua-privy tank. There were design faults with the settled sewer network, principally the shallow (often zero) depth of the connector and collector sewers and their junction boxes, but the system has worked well hydraulically.

7.4 CONCLUSIONS

Settled sewerage schemes installed in Africa have been mainly sewered aqua-privies. However, the aqua-privy itself is a questionable sanitation technology: the water-tight tank is expensive, and a better lower-cost alternative is the pour-flush toilet, which can of course be sewered (Kalbermatten *et al.*, 1983) (and as recommended for parts of Tunis: Mara, 1992). The installation of sewered aqua-privies *de novo* thus cannot now be recommended, and a decision as to whether to install settled sewerage should follow the rationale set out by Mara (Chapter 2). However, the development of the sewered aqua-privy system by Vincent *et al.* (1963) has been most valuable as it has provided tropical, and indeed temperate, public health engineers with the now well proven technology of settled sewerage.

7.5 REFERENCES

Black and Veatch International (1975). *Storm Drainage and Sanitary Sewerage Master Plan Report for The City of N'Djamena, Chad.* Abidjan: African Development Bank.

de Kruijff, G.J.W. (1978). *Aqua-Privy Sewerage Systems: A Survey of Some Schemes in Zambia.* Nairobi: University of Nairobi (Housing Research and Development Unit).

Feachem, R.G., Mara, D.D. and Iwugo, K.O. (1978a). *Sanitation Studies in Africa—Site Report No. 4: Zambia (Lusaka and Ndola).* Unpublished research report. Washington, DC: The World Bank.

Feachem, R.G., Mara, D.D. and Iwugo, K.O. (1978b). *Sanitation Studies in Africa—Site Report No. 2: Nigeria (New Bussa).* Unpublished research report. Washington, DC: The World Bank.

Feachem, R.G., Mara, D.D. and Iwugo, K.O. (1979). *Alternative*

Sanitation Technologies for Urban Areas in Africa. P.U. Report No. RES 22. Washington, DC: The World Bank.

Kalbermatten, J.M., Julius, D.S., Gunnerson, C.G. and Mara, D.D. (1983). *Appropriate Sanitation Alternatives: A Planning and Design Manual*. Baltimore: Johns Hopkins University Press.

Mara, D.D. (1976). *Sewage Treatment in Hot Climates*. Chichester: John Wiley & Sons.

Mara, D.D. (1992). *Solutions Alternatives d'Assainissement Appropriées au Grand Tunis*. Report for Greater Tunis Sewerage Masterplan Study. Leeds: Lagoon Technology International.

Martin, A.J. (1935). *The Work of the Sanitary Engineer*. London: Macdonald and Evans.

National Housing Authority (1979). *Urban Sanitation Survey*. Lusaka: NHA.

Otis, R.J. and Mara, D.D. (1985). *The Design of Small Bore Sewer Systems*. TAG Technical Note No. 14. Washington, DC: The World Bank.

Reed, R. and Vines, M. (1991). *Evaluation of Sewered Aqua-privies in Kabushi, Ndola, Zambia* Loughborough: University of Technology (WEDC).

Reed, S. and Vines, M. (1992). *Sewered Aqua-privies in New Bussa, Kwara State, Nigeria*. Loughborough: University of Technology (WEDC).

Smith, F. (1993). *Guidelines for the Design, Operation and Maintenance of Septic Tank Effluent Drainage Systems in South Africa, with reference to the Marselle Case Study*. Research Report No. 700. Pretoria: CSIR.

Vincent, L.J., Algie, W.E. and Marais, G.v.R. (1963). A system of sanitation for low cost high density housing. In *Proceedings of the Symposium on Hygiene and Sanitation in Relation to Housing CCTA/ WHO, Naimey 1961*, Publication No. 84, pp. 135–172. London: Commission for Technical Cooperation in Africa South of the Sahara.

Vines, M. and Reed, R. (1991). *Evaluation of Sewered Aqua-privies in Matero, Lusaka, Zambia*. Loughborough: University of Technology (WEDC).

Waddy, B.B. (1971). The siting and sewage system of New Bussa. *Journal of the Society of Health of Nigeria*, **6**(1), 16–19.

8

Hydraulic Design of Unconventional Sewerage

Martin J. Marriott

8.1 INTRODUCTION

As a preface to the discussion of *un*conventional systems, there should be a definition of what is meant by a conventional system. Conventional sewerage is taken to be that which serves the major cities and towns of most European countries. With some exceptions, such as post-war new towns in the UK, these systems have evolved and developed over many years. An historical perspective is therefore beneficial, to understand why and how the systems were designed. As all engineering and technology should be applied in an appropriate way, it is worth examining the roots of conventional practice, to determine which criteria remain appropriate in other contexts. With this in mind, conventional practice in the UK is described, with some initial comments on the historical development of sewerage in London.

8.2 HISTORICAL DISCUSSION

The concept of a foul sewer is a relatively modern idea. The term sewer was originally derived from land drainage. Many of London's main sewers originated as culverted water courses that initially carried only surface water. Households were served by cesspools, or might simply deposit their waste in the streets. Originally it was an offence to discharge waste to the 'river sewers', but later practice changed. Individual cesspools were abandoned, and many dwellings were connected to the sewers, which at the time drained directly into the River Thames. Conti-

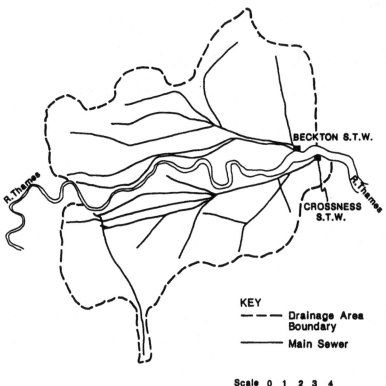

KEY

— — — Drainage Area
            Boundary

———— Main Sewer

Scale 0 1 2 3 4
          |__|__|__|__|
              miles

Figure 8.1 Plan of London's sewerage, showing the main sewers and the drainage areas of Beckton and Crossness sewage treatment works

nuing outbreaks of cholera, and the foul smelling river in the 1850s, led to the implementation of a sewerage scheme designed by Sir Joseph Bazalgette. Main interceptor sewers were constructed to collect flows which had previously entered the Thames in central London, and to convey the sewage east to outfalls at Barking and Crossness, some distance downstream of the city centre. This remains the basis of the London sewerage system to this day, as shown in Figure 8.1.

Work on Bazalgette's scheme for London was completed in the 1870s, but it was not until the late 1880s that any treatment was provided at the outfall sites. Worth and Crimp (1897) described the construction and operation of precipitation channels at Barking and Crossness, and the commissioning of sludge vessels to dump the resulting sludge out at sea. The extent of treatment provided was later increased to include biological stages. The latest development, resulting from the ban on sludge dumping to sea from 1998, is the planned provision of facilities for sludge incineration.

Bazalgette's original design data for London may be summarized as follows:

Population	3.45 million
Sewage	108 million gallons (23 700 m^3) per day (24 hours)
Rainfall	286 million gallons (62 900 m^3) per day (24 hours)

The sewers were constructed to carry one-half of that total quantity in 6 hours of the day. In other words, a peaking factor of 2 was allowed. Rainfall in excess of a quarter of an inch per day (approximately 6 mm/day) was diverted to the river. It was estimated at the time that the capacity of the sewers would not be exceeded on more than 12 days in a year. The system was designed as combined sewerage on cost grounds, although there were those who would have favoured separation of the foul and surface water flows.

At the turn of the century, a wide ranging presidential address to the Institution of Civil Engineers (Hawksley, 1901) included comment on the separate system of drainage, which was considered not only costly, but also 'in most cases undesirable, inasmuch as the washings from the surfaces of the streets must necessarily be of a foul nature so long as animal traction remains in use, whatever may be the case when, if ever, the employment of horses is entirely superseded by mechanical motors'.

With rapid and continuous growth of towns and cities in the UK, partially separate systems were developed which allowed the flow from back roofs and yards to be collected with the domestic sewage, and for storm water from other areas to be collected separately. Lloyd-Davies (1906) described improved methods of calculating such flows, and a Mr Fowler in discussion on that paper mentioned serious problems with flooding in Leeds. He advocated the use of leap-weirs to separate a proportion of the flood flow, and mentioned the benefits of traditional egg-shaped sewers, to give a better dry-weather-flow velocity. The type of overflow structure he described is illustrated, although no longer recommended, in a recent guide to the design of combined sewer overflow structures (Balmforth *et al.*, 1994), and it is noted that precast concrete ovoid sections complying with current standards have recently become commercially available in the UK.

8.3 CONVENTIONAL DESIGN

Current conventional UK sewerage design is covered by the
British Standard 8005 (BSI, 1987). Many points are also usefully
summarized in the design and construction guide for developers
in England and Wales, entitled *Sewers for Adoption* (WSA,
1995). The separate system is normally adopted for new development in the UK, unless the extent of existing combined or
partially separate sewers determines otherwise. The hydraulic
design of a combined system is more closely associated with
surface water sewerage, and it is not intended to cover that in
detail in this chapter. The main points that relate to the
hydraulic design of conventional foul sewers are discussed below.

8.3.1 Design flow

In the absence of specific local water use data, a figure of 200
litres per person per day plus 10 per cent infiltration may be
assumed. This is multiplied by the population appropriate to the
planning horizon adopted (30 years is recommended) to give the
average flow or dry weather flow (DWF). Foul sewers are
designed for up to 6 × DWF, to allow for diurnal peaks, other
fluctuations and extraneous flow such as infiltration. With the
assumption of three persons per dwelling, this approximates to a
peak design flow of 4000 litres per dwelling per day or, in more
meaningful units for a peak flow, 0.046 litres per second per
dwelling.

8.3.2 Pipe size, depth and roughness

Sewers laid within highways should have a minimum cover of 1.2
m to avoid interference with other services. Elsewhere, a
minimum cover of 0.9 m is suggested as sufficient to avoid interference with cultivation of the land. The minimum size of gravity
sewer for adoption into the public system is 150 mm internal
diameter. House connections are usually 100 mm internal
diameter, thus ensuring that solids which enter the system can
pass downstream through pipes of increasing size to prevent
blockages. A surface roughness height (k_s) value of 1.5 mm for

use in the Colebrook–White equation is usually now assumed for design purposes. This is approximately equivalent to a Manning's *n* of 0.013, which is close to the value of 0.012 that equates to the traditional Crimp and Bruges formula.

8.3.3 Velocity

Sewers should be laid at gradients that will produce velocities sufficient to prevent permanent deposits of solids. A velocity of 0.75 m/s should be exceeded daily in foul sewers to maintain self-cleansing conditions. The *Sewers for Adoption* guide requires this velocity at one-third of the design flow, that is at 2 × DWF, which represents a typical diurnal peak DWF. BS8005 includes the comment that this can be achieved by laying the sewers to a gradient that will give a velocity of 1.0 m/s at full bore flow. This allows for the reduction in velocity in pipes flowing less than half full.

Sewers for Adoption also includes a 'deemed to satisfy' criterion whereby, provided 10 dwellings are connected, the minimum velocity requirement is deemed to be satisfied by a 150 mm diameter sewer having a gradient not flatter than 1 in 150. This is to limit the extent of pumping required in flat regions, but is not advocated where the topography permits steeper gradients.

8.3.4 The self-cleansing criterion

It has been shown from sediment transport criteria (Ackers, 1984) that the British Standard recommended self-cleansing velocity appears inadequate in pipes of large diameter. This would apply to the large collector culverts, many several metres in diameter, employed in the Greater Cairo Wastewater Project (Darling and Drake, 1985). In contrast, Novak and Nalluri (1978) concluded, particularly for smaller sewers less than 1000 mm in diameter, that the standard self-cleansing velocity results in greater slopes than would be required to initiate motion even for large grit and sediment particle sizes, when considering flow over a fixed bed.

More recent research into the nature of sewer sediments (Crabtree, 1989), has shown that a distinction should be made between the high bed shear stresses required to erode the consolidated cohesive deposits found particularly in combined sewers (which in many cases exceeded 800 N/m^2), and the relatively lower values that relate to the threshold of motion of freshly deposited sanitary solids.

It may be demonstrated (Marriott, 1994) that threshold-of-motion values reported for recently deposited solids are similar to the average boundary shear stress (τ) experienced by a pipe of internal diameter D (mm) laid at a gradient of 1 in D and flowing full:

$$\tau = \rho g(D/4)\,(0.001/D)$$
$$= 2.5 \text{ N/m}^2$$

This unit tractive force or boundary shear stress approach has been used in some standards, for example the Metropolitan Water Sewerage and Drainage Board, Sydney (1979), where limiting gradients (S, %) were specified as:

$$S = (0.0135 \,/\, R) \text{ for self-cleansing}$$

$$S = (0.034 \,/\, R) \text{ for sulphide slime control}$$

where R = hydraulic radius (m).

The slope of a pipe flowing full with gradient 1 in D may be seen to lie between these two values, with $S = (0.025 \,/\, R)$. The 'deemed to satisfy' criterion from *Sewers for Adoption* for low flows is a form of this gradient rule, although the average boundary shear stress will be reduced for flows and proportional depths below half full. With respect to low flows, it is worth mentioning that the effective peaking factor from a small number of dwellings is likely to be rather higher than the values traditionally assumed for sewer design. We are here at the interface between building drainage and sewerage. Butler (1991) recorded a mean flow rate of 1.0 litres per second from WC cisterns, in a survey of discharges from domestic appliances. This is over 20 times the peak design flow per dwelling of 0.046 litres per second. There will be some attenuation of short duration discharges, but clearly the peaks inherent in appliance discharges will be beneficial for self-cleansing. Notwithstanding this, the 'deemed to satisfy' low flow criterion is still considered to be a compromise to avoid excessive depth and pumping, and not necessarily to achieve a self-cleansing regime. Where topography permits, steeper gradients are preferred in conventional practice.

8.3.5 Improved separate systems and source control

Developments in sewerage systems in Europe are likely to reflect the requirements of the Urban Waste Water Treatment Directive. Discharges from combined sewer overflows, together with the polluted nature of surface water runoff in urban areas, often produce an adverse environmental impact on water courses. Construction Industry Research and Information Association (1992) describes scope for control of urban runoff, by methods such as infiltration techniques. Proposals in the Netherlands include an 'improved' separate system, whereby the more polluted portion of flow from the stormwater system is accepted into the foul system for eventual full treatment. These developments often appear more costly to the individual householder or developer, unless considered in an overall environmental sense.

8.4 UNCONVENTIONAL SEWERAGE

The review of conventional sewerage above shows how practice has evolved, and is likely to continue to develop, both at the level of overall concepts, and concerning detailed design criteria. Similarly, there may be the need for practice to be modified in different situations. Cotton and Franceys (1991) discuss the provision of infrastructure for low-income sites in developing countries, and report successful sewerage design using a minimum velocity criterion of 0.5 m/s. Pickford (1995) summarizes unconventional methods of sewerage whereby a saving in capital cost has been achieved, by one or more of the following:

- using smaller diameter pipes,

- laying pipes at flatter gradients;

- laying sewers at shallow depth;

- laying sewers within plots at the rear of premises, and

- providing interceptor tanks for settlement of solids

Other chapters in this book illustrate to what extent conventional criteria may be simplified and reduced, and with what effect. However, the following approaches have distinct differences from conventional systems, and the hydraulic criteria that apply are summarized.

8.4.1 Settled sewerage

In this system, also known as small bore sewerage, domestic sewage flows through an interceptor tank prior to discharge to the sewer. This means that the sewerage system receives only the liquid portion of the sewage for conveyance. Solids are intended to accumulate in the interceptor tank for periodical removal and disposal. Otis and Mara (1985) describe many practical considerations for effective operation of the interceptor tank. Provided the tank operates as intended, the sewer may be laid at a nominal fall, and may include sections of pressure flow where the sewer drops below the hydraulic grade line. In that respect, the hydraulic design becomes like that for gravity water supply pipelines, which may follow the ground contours, provided the point of discharge is at a lower elevation than the point at which flow enters the system, and consideration is given to the possibility of air locks.

Unless specialized cleaning equipment is available, a minimum pipe size of 100 mm was recommended, with periodic flushing from manholes to create a flow velocity of at least 0.5 m/s. With solids removed from the sewage, there is no requirement for a daily self-cleansing velocity. With considerable attenuation of individual peak discharges by the interceptor tanks, a reduced design peaking factor of 2 was recommended pending the availability of more field data. Flat gradients may be accommodated by allowing the pipeline to follow within reason the natural ground contours, in what is referred to as the inflective gradient design approach.

An example quoted was the sewered septic tanks or common effluent drains in South Australia. These arise where during the initial development, houses are served by septic tanks and soakaways. At a later stage, these septic tanks are connected to a small bore sewerage system, with economy over a conventional installation. Design data quoted from South Australia were as follows:

Minimum pipe size	100 mm
Average flow per person	136 litres per day
Peaking factor	3 times the average dry weather flow
Minimum velocity	0.46 m/s
Minimum slope	100 mm diameter at 1 in 150
	150 mm diameter at 1 in 250
	200 mm diameter at 1 in 300

Pipes were required not to flow more than half full at any time. Annual flushing was recommended. It is understood that these systems operate very satisfactorily. The saving in construction cost relates to the use of smaller and shallower pipes than for conventional sewerage. However, the velocity and gradient requirements quoted are still close to conventional sewerage criteria, and do not take full advantage of the solids-free nature of the sewage. The required half full limit, appears similar to the UK building drainage requirement (BS8301: BSI, 1985) that the peak flow in a foul drain should not exceed a proportional depth of 0.75 to ensure adequate air flow and avoid the risk of induced trap siphonage. However, this conflicts with the possible use of the inflective gradient approach.

8.4.2 Vacuum sewerage

In some flat areas of the UK, where provision of conventional gravity sewers is difficult and costly, vacuum systems have been employed. Stanley and Mills (1984) show cost savings of over 20 per cent, while others (Ashlin *et al.*, 1991) quote more modest savings. Satisfactory operation relies on effective elimination of surface water flows.

Although these results appear to show a promising application of unconventional sewerage particularly in flat regions, and also perhaps where water is scarce, the relatively small cost reductions do not put vacuum sewerage in the low-cost category.

Typical design criteria used for vacuum sewerage are as follows:

Minimum pipe size   75 mm

Minimum gradient   0.2 per cent in flow direction, with regular lifts

Cover   1.0–1.5 m

Flow   Similar to conventional UK sewerage

Velocity   The sewage may be conveyed at velocities of about 5 m/s to the vacuum collection vessel.

8.5 CONCLUSIONS

In developed countries, the original benefits to public health from a comprehensive sewerage system may often be taken for granted. Public concern has perhaps shifted towards wider environmental considerations. Operations have become automated, and work forces reduced. Performance criteria for design and operation reflect a balance between convenience, control and cost that is judged to be acceptable and affordable to the customers. As these design criteria for foul sewerage have evolved to suit this context, it is argued that they should not necessarily be mindlessly transferred to other situations.

Conventional hydraulic criteria for foul sewerage have been described and discussed, with a historical perspective from the UK. It has been demonstrated that the conventionally adopted self-cleansing velocity does not reflect a single absolute theoretical value, but that conditions for sediment movement depend upon the type of sediment and sewer.

The hydraulic design of settled or solids-free sewerage has been contrasted with conventional design. The resulting use where appropriate of smaller, shallower pipes that may follow the natural ground contours, should offer scope for providing the main benefits of effective sanitation at a considerably lower cost than conventional sewerage.

8.6 REFERENCES

Ackers, P. (1984). Sediment transport in sewers and the design implications. In *Proceedings of the International Conference on the Planning, Construction, Maintenance and Operation of Sewerage Systems*, pp. 215–230. British Hydraulics Research Association: Cranfield.

Ashlin, D.E., Bentley, S.E. and Consterdine, J.P. (1991). Vacuum sewerage: the Four Crosses experience. *Journal of the Institution of Water and Environmental Management*, **5** (6), 631–640.

Balmforth, D.J., Saul, A.J. and Clifforde, I.T. (1994). *Guide to the Design of Combined Sewer Overflow Structures*. Report No.FR0488. Swindon: Foundation for Water Research.

British Standards Institution (1985). *Building Drainage BS8301: 1985*. London: BSI.

British Standards Institution (1987). *Sewerage. BS8005: Part 1:1987*. London: BSI.

Butler, D. (1991). A small-scale study of wastewater discharges from domestic appliances. *Journal of the Institution of Water and Environmental Management*, **5** (2), 178–185.

Construction Industry Research and Information Association (1992). *Scope for Control of Urban Runoff*. Report Nos 123 and 124. London: CIRIA.

Cotton, A. and Franceys, R. (1991). *Services for Shelter*. Liverpool: Liverpool University Press.

Crabtree, R.W. (1989). Sediments in sewers. *Journal of the Institution of Water and Environmental Management*, **3** (6), 569–578.

Darling, R.S. and Drake, J. (1985). Greater Cairo wastewater project: studies and master plan. *Proceedings of the Institution of Civil Engineers (Part 1)*, **78**, 745–763.

Hawksley, C. (1901). Presidential address. *Minutes of the Proceedings of the Institution of Civil Engineers*, **147**, 2–62.

Lloyd-Davies, D.E. (1906). The elimination of storm-water from sewerage systems. *Minutes of the Proceedings of the Institution of Civil Engineers*, **164**, 41–195.

Marriott, M.J. (1994). Self-cleansing sewer gradients. *Journal of the Institution of Water and Environmental Management*, **8** (4), 360–361.

Metropolitan Water Sewerage and Drainage Board (1979). *Design of Separate Sewerage Systems*. Sydney: MWSDB.

Novak, P. and Nalluri, C. (1978). Sewer design for no sediment deposition. *Proceedings of the Institution of Civil Engineers (Part 2)*, **65**, 669–674.

Otis, R.J. and Mara, D.D. (1985). *The Design of Small Bore Sewer Systems*. TAG Technical Note 14. Washington, DC: The World Bank.

Pickford, J. (1995). *Low Cost Sanitation—A Survey of Practical Experience*. London: IT Publications.

Stanley, G.J. and Mills, D. (1984). Vacuum sewerage. In *Proceedings of the International Conference on the Planning, Construction, Maintenance and Operation of Sewerage Systems*, pp. 317-326. BHRA, The Fluid Engineering Centre, Cranfield, Bedford.

Water Services Association (1995). *Sewers for Adoption*, 4th edn. Swindon: Water Research Centre.

Worth, J.E. and Crimp, W.S. (1897). The main drainage of London. *Minutes of the Proceedings of the Institution of Civil Engineers*, **129**, 49–187.

9

Small Diameter Gravity Sewers: Experience in the United States

Richard J. Otis

9.1 INTRODUCTION

Water pollution and public health problems from inadequately treated wastewater exist in many small communities throughout the USA. Failures of individual septic tank systems have forced many unsewered communities to consider construction of public collection and treatment facilities. Often, severe financial hardships are created in providing these facilities. It is not uncommon for residents of small communities to pay two to three times as much for sewer services as residents of larger municipalities (Graham, 1982). This higher cost is due primarily to the length of sewer per user which can be more than five times longer than that typical in urban areas. The impact of the higher cost on family budgets can be quite severe because the average annual incomes in rural communities are significantly lower than in more urbanized areas. As a result, plans for construction of needed public facilities are often rejected, allowing public health hazards, nuisances and environmental degradation from failing septic tank systems to continue while economic development is impeded.

Traditionally, conventional sewerage is assumed to be necessary without consideration of alternatives. The cost-effectiveness of various treatment alternatives are commonly evaluated, but while the extent of the sewers is sometimes a cost issue, low-cost alternatives for wastewater collection are rarely considered. Yet, conventional sewer construction accounts for 80–90 per cent of the total capital costs of the collection and treatment facility (Kreissl *et al.,* 1978) and more than 65 per cent of the total annual costs (Smith and Eilers, 1970). Less costly, but equally effective alternatives to conventional wastewater collection systems have been developed and have been shown to significantly reduce wastewater facility costs. Only recently have these alternatives been seriously considered.

Development of less-costly collection alternatives has centered around changing the motive force and/or the character of the wastes collected to reduce excavation and pipe costs. Table 9.1 lists the most common alternatives. Small diameter gravity sewers (SDGS) were the most recent to be introduced, but rapidly gaining in popularity. They have been called various names including variable grade sewers (Simmons *et al.,* 1982), small bore sewers (Otis and Mara, 1985), and common effluent drains (South Australia Health Commission, 1986). They are all similar in design. Unlike conventional sewers, SDGS collect settled wastewater. Grit, grease and other troublesome solids which might cause obstructions are separated from the waste flow in septic or interceptor tanks upstream of each connection. The solids which accumulate in the tanks are removed periodically for treatment and disposal.

Although the term 'small diameter gravity sewer' has become commonly accepted, it is not an accurate description of the system. The mains need not be small in diameter (the size is determined by hydraulic considerations), nor are they 'sewers' in the sense that they carry wastewater solids. The most significant feature of SDGS is that primary pretreatment is provided in interceptor tanks upstream of each connection. This offers several advantages over conventional sewers. With the settleable solids and grease removed, it is not necessary to design the collector mains to maintain minimum self-cleansing velocities. Without the requirement for minimum velocities, the pipe gradients may be reduced and, as a result, excavation depths and the number of lift stations are also reduced. The need for manholes at all junctions, changes in grade and alignment, and at regular intervals is eliminated. The interceptor tank attenuates the wastewater flow rate from each connection which reduces the peak-to-average flow ratio below what is typically used for establishing design flows for conventional sewers. Without the requirement to carry solids, the depth of flow in the sewer is not critical. Therefore, SDGS can be employed without concern in areas

Table 9.1 Alternative wastewater collection systems

Type	Description	Motive force	Waste character
Grinder pump pressure sewers (GP)	Small vaults at each connection store wastewater for periodic pumping into small pressure mains (min. dia. 30 mm) laid with uniform depth of burial	Positive pressure provided by individual pumping units	Ground fresh waste, which may become septic in extensive systems
Septic tank effluent pumping pressure sewers (STEP)	Interceptor tanks at each connection remove settleable solids and provide storage of settled waste for periodic pumping in to small pressure mains (min. dia. 30 mm) laid with uniform depth of burial	Positive pressure provided by individual pumping units	Septic settled waste
Vacuum sewers	Pneumatic valves at each connection release small volumes of wastewater and air into vacuum mains laid at shallow depth with 'sawtooth' pattern	Negative pressure provided by central vacuum pump station	Fresh oxygenated raw waste
Small diameter gravity sewers (SDGS)	Septic tank effluent free from grit, grease and large solids is collected in small mains (min. dia. 50 mm) not designed for self-cleansing velocities	Gravity	Septic settled waste

where severe water conservation is practised or where long flat reaches are laid with few connections. Also, headworks (screens and grit removal) and primary sedimentation can be eliminated from the treatment plant as these unit processes are performed by the interceptor tanks at each connection. In addition, sludge volume and treatment are reduced because the interceptor tanks provide sludge digestion. Furthermore, as the sludges are segregated at each connection, they remain uncontaminated by commercial or industrial sludges, so reducing restrictions on sludge disposal.

9.2 HISTORICAL DEVELOPMENT AND USE

Late in the 1950s, severe problems with surface seepage of wastes from failing septic tank systems in several rural subdivisions surrounding the city of Adelaide in South Australia prompted the Department of Public Health to take corrective action. Three solutions were proposed: (a) restricted water use, (b) cartage, and (c) sewers (South Australia Department of Health, 1968). Restrictions on water use proved to be unsuccessful and the costs of cartage were prohibitive. However, conventional sewers also were too costly. As an alternative, 'common effluent schemes' were recommended based on limited experience in seaside towns. These schemes were small diameter 'drains' used to collect septic tank effluent from the existing septic tanks for subsequent treatment in oxidation ponds. They were designed with continuous negative gradients similar to conventional sewers. Cost-saving features were smaller conduit size, reduced pipe gradients, fewer manholes, lower pumping costs, and less operation and maintenance (South Australia Department of Health, 1968; Gutteridge, Haskins and Davey Ltd, 1977).

The first scheme was constructed in Pinnaroo, South Australia, in 1962. By 1986, over 80 schemes had been constructed. The largest scheme serves 4000 connections (South Australia Health Commission, 1986). Despite the lack of routine main flushing, obstructions have been infrequent. Where they have occurred, they were most often due to construction debris left in the conduits after installation.

Small diameter gravity sewers were not introduced in the USA until the mid-1970s. The first systems were in Mt Andrew, Alabama and Westboro, WI, which were small demonstration systems of 13 and 90 connections, respectively (Otis, 1986). The Mt Andrew system was constructed as a variable grade system with sections of sewer depressed below the static hydraulic grade line (Simmons *et al.*, 1982). The Westboro system was designed with uniform gradients using the more conservative Australian guidelines (Otis, 1978).

Since the early 1980s the use of SDGS has rapidly increased in the USA primarily in existing rural communities. By 1990, over 200 systems were operating. More recently, they have been used increasingly for residential fringe developments, new subdivisions, and resort developments where the topography is favorable. Frequently, the systems built are hybrid gravity and pressure systems (Otis and Sirotiak, 1987). Experience with SDGS has been excellent and, as a result, the designs have become less conservative.

Unfortunately, there have been significant deterrents to their use from regulatory agencies, engineers, and the consumers. Engineers and regulatory agencies have been reluctant to promote SDGS because of the concern for long-term performance. Potential users have fought against their construction because of a perception that they are 'second rate'. Typical concerns are over backups, odours and potential for expansion. It has been proven, however, that these concerns can be overcome through proper planning and design. With time, it is expected that SDGS will receive full acceptance.

9.3 DESIGN GUIDELINES

SDGS design practices have rapidly evolved in the USA as experience has been gained. The guidelines originally imported from Australia have gradually been modified to allow conduits as small as 50 mm in diameter, variable pipe gradients, depressed sections (below the hydraulic grade line), elimination of minimum flow velocities and reduction in peak-to-average flow factors.

9.3.1 Estimation of design flow

The collector mains are sized to carry the maximum daily peak flow. Conventional sewer design typically assumes 380 litres per caput per day times a peaking factor of 4 to establish the necessary collector capacity. This estimate includes allowances for commercial flows and infiltration. Experience with SDGS has shown this design flow criterion is excessive because of the capacity of the interceptor tanks to attenuate the wastewater peaks. In residential dwellings, the rate of wastewater discharged from the building depends on the type of water fixtures and appliances used. Instantaneous peak flows may be 0.3–0.6 l/s. However, the interceptor tank can attenuate the peaks dramatically. Monitoring of individual interceptor tanks showed that outlet flows ranged between 0.03 and 0.06 l/s over periods of 30–60 minutes (Otis, 1986). Long periods of zero flow also occur. The extent of attenuation depends on the design of the interceptor tanks and its outlet (Jones, 1975). Most recent SDGS designs have been based on design flows of 0.006–0.02 l/s per connection. Designs based on these flow criteria have been successful where watertightness of all tanks and piping is thoroughly tested during construction. Also, storage is available in the interceptor tank above the normal water level for household flows for short peak flow periods if they should occur (EPA, 1991).

9.3.2 Hydraulic design

Hydraulic computations for SDGS are the same as those used for conventional gravity sewers. Typically, Manning's pipe flow formula is used with a roughness coefficient (n) of 0.013. However, unlike conventional sewers, SDGS can alternate between open channel and pressure flow. Therefore, separate analyses must be made for each segment of the sewer in which the type of flow does not change. Where pressure flow occurs, the elevation of the energy grade line during daily peak flow conditions must be calculated to determine that no interceptor tank outlet invert lies below the grade line. If not, free-flowing conditions will not be maintained at those connections. Where this is determined to occur, the designer has several options: (a) the elevation of the downstream summit in the sewer can be lowered to lower the grade line; (b) the diameter of the main can be increased to reduce the frictional headloss and velocity head; or (c) a residential lift station or STEP unit can be installed to lift the wastewater into the collector. If short term surcharging above any interceptor tank outlet is expected, check valves on the individual service lateral may be sufficient to prevent backflow.

Experience has shown that the normal flows that occur in the systems are sufficient to keep the mains solids-free. The solids that enter the collectors from the interceptor tanks and the slimes which may grow within the sewer are easily carried when flow velocities of 0.15 m/s are achieved intermittently (Otis, 1986). Thus, SDGS need not be designed to maintain minimum flow velocities during peak flows, although many state agencies require that minimum velocities of 0.3–0.45 m/s be reached daily.

The pipe diameter is determined by hydraulic analysis. The minimum diameter is typically 100 mm, but 50 mm has been used successfully. Where 50 mm pipe is used, flow control devices connected to the interceptor tank outlet to limit peak flows and check valves installed in the service lateral to prevent backflow are used. The costs of the flow control devices and check valves generally cancel savings realized from the smaller pipe and, therefore, 100 mm pipe is most commonly used.

9.3.3 Collector mains

The collectors are laid out in a typical dendriform pattern similar to that of conventional sewers. Because installation is substantially less costly, mains are frequently run on both sides of a paved roadway to reduce the costs associated with pavement restoration. Alternatively, in many communities, the collectors are laid in the alleys behind the homes as the existing septic tank systems are usually located in the back of the lots.

Strict horizontal and vertical control during construction has not been necessary. Although utilities, large trees, rock outcroppings, etc. are usually avoided with careful planning, unforeseen obstacles often occur, but may be avoided during construction by bending or rerouting the pipe. Only elevations of major summits need to be controlled to ensure that upstream connections are not flooded. Where depressed sections occur, air venting is necessary at the summits. Between summits, the profile of the sewer should be reasonably uniform so unvented air pockets do not form to create unanticipated headlosses and excessive upstream surcharging.

In most systems, cleanouts are substituted for manholes except at major junctions. Cleanouts are one-fifth the cost of manholes, provide sufficient access for hydraulic flushing, and are not a source of infiltration, inflow, or grit. Manholes have also been a source of odours in SDGS systems.

9.3.4 House connections

Prefabricated, single compartment septic tanks are usually used for interceptor tanks. For residential connections, tank volumes of 3.4–4.0 m^3 are used. For commercial connections, tank volumes are usually determined based on local septic tank system codes.

Watertightness of the tanks is critical. For this reason, existing septic tanks are usually replaced with new tanks including a new building sewer. Projects in which existing tanks have been used havé experienced excessive wet weather flows. Vacuum testing is often used to test watertightness. The typical acceptance criterion is less than 2.5 cm loss of mercury vacuum after 5 minutes with an initial vacuum of 10 cm. The test is run after connections are made and include the manholes.

In some SDGS systems, modifications have been made to the tank outlet to control maximum flow rates. This has been advocated to allow smaller diameter pipe to be used for the collection mains. Surge tanks installed after the interceptor tanks were used in the past, but they created excessive head loss in the system requiring unnecessary excavation and odour problems. Flow control devices are available which can be mounted inside the interceptor tank to maintain more uniform flow discharges. The freeboard available in the tank is used for the necessary storage. Most SDGS systems, however, have not used flow control devices.

9.3.5 Odour and corrosion problems

Odours and corrosion are the most commonly reported problems associated with SDGS. Odours typically occur at lift stations, air release valves and house plumbing stack vents. Odours are most pronounced where turbulence occurs to release the dissolved gases in the settled septic waste. Odours at lift stations are often successfully abated by installing drop inlets that extend below the pump off level. Drop inlets eliminate most of the turbulence. Other successful control measures include soil odour filters, and airtight wet well covers with vents that extend 3–5 m above grade. Soil odour filters are the usual method to eliminate odors from air release valves.

Odours at individual connections usually originate from turbulence in the collection main, most often on steep runs where hydraulic jumps occur at a change in slope. The odours which are released flow back up the sewer due to the 'chimney' effect and exit at the stack vents of the connections at the highest elevation. Successful corrective measures have included placing running traps in the service laterals or extending the main further uphill and terminating it with an air vent and soil odour filter.

Corrosion is most severe at lift stations where turbulence occurs to release hydrogen sulphide in the wastewater. Lift station equipment and appurtenances must be carefully selected to be corrosion resistant and drop inlets used to reduce turbulence.

9.4 MANAGEMENT

Typically, utility or special purpose districts are formed to administer, operate, and maintain SDGS systems located outside municipal boundaries. These districts vary in structure from state to state, but they usually have most of the powers of municipal government, except for bonding authority.

The districts usually assume responsibility for all SDGS components downstream from the inlet to the interceptor tank. As the interceptor tank is critical to the performance of the system, it is important that the utility has the responsibility for its maintenance. In most projects, the utility owns the interceptor tank and service lateral. A general easement, or easement by exhibit, has worked well to gain access to the components located on private property.

Septage pumping, treatment and disposal are usually provided by private pumpers under contract to the district. Pumping frequencies vary depending on the type of connection. Food service establishments may need pumping as often as every 6 months while residential homes may not need pumping for 7–10 years.

Mains flushing usually is recommended on an annual basis, but experience in Australia and the USA has shown that routine flushing is not necessary. Successful hydraulic flushing can be accomplished by discharging clear water into cleanouts and removing the accumulated solids from downstream manholes or lift station wet wells. Flushing two to three pipe volumes has been shown to be effective in removing all slimes growths from the pipe wall (Otis, 1986). However, flushing has seldom, if ever, been performed in SDGS systems in the USA and Australia.

9.5 CONSTRUCTION COSTS

A review of construction costs from 12 SDGS projects showed that 20–50 per cent savings were achieved over conventional gravity sewers, with an average savings of 30 per cent (Otis, 1986). The average cost per connection was US$5,353 (1991). Although the costs per connection were higher than is typical for urban areas, the installation cost per meter of pipe was significantly less than that for conventional sewers. Of the projects reviewed, the in-place pipe costs averaged US$49.51 with a range of US$8.75–56.72. The average total project cost per metre of collection main installed was US$189.80 (EPA, 1991). The installation of the collectors and interceptor tanks accounted for over 50 percent of the total costs of construction. In projects where the collectors could have shallow burial, and cleanouts rather than manholes were used, the construction costs per meter of pipe installed were the lowest of the projects reviewed.

9.6 SUMMARY

SDGS have potential for wide application. They are a viable alternative to conventional sewers in many situations, but are particularly well suited for low-density residential and commercial developments such as small communities and residential fringe developments of larger urban areas. Because of their smaller size, reduced gradients and fewer manholes, they have a distinct cost advantage over conventional gravity sewers where adverse soil or rock conditions create mainline excavation problems or where restoration costs in developed areas can be excessive. In new developments, construction of the sewers can be deferred until the number of homes built warrant their installation. In the interim, septic tank systems or holding tanks can be used. When the sewers are constructed, the tanks can be converted for use as interceptor tanks. However, SDGS usually are not well suited in high-density developments because of the cost of installing and maintaining the interceptor tanks.

9.7 REFERENCES

EPA (1991). *Alternative Wastewater Systems*. Report No. EPA/625/1-91/024. Washington, DC: Office of Research and Development, Office of Water, Environmental Protection Agency.
Gutteridge, Haskins and Davey Ltd. (1977). *Planning for the Use Of*

Sewage: Summary Report. Canberra: Australian Government Publishing Services.

Graham, M.J. (1982). Study shows variety in sewer charges. *Deeds and Data*, **19**(1), 12. Arlington,VA: Water Pollution Control Federation.

Jones, E.E., Jr. (1975). Domestic water use in individual homes and hydraulic loading of and discharge from septic tanks. In *Proceedings of the National Home Sewage Disposal Symposium* (Conrad Hilton Hotel, Chicago, 9–10 December 1974), pp. 89–103. St Joseph, MI: American Society of Agricultural Engineers.

Kreissl, J.F., Smith, R. and Heidman, J.A. (1978). *The Cost of Small Community Wastewater Alternatives: Training Seminar for Wastewater Alternatives for Small Communities*. Cincinnati, OH: Municipal Engineering Research Laboratory, Environmental Protection Agency.

Otis, R.J. (1978). *An Alternative Public Wastewater Facility for a Small Rural Community*. Madison, WI: Small Scale Waste Management Project, University of Wisconsin.

Otis, R.J. and Mara, D. (1985). *The Design of Small Bore Sewer Systems*. TAG Technical Note No. 14. Washington, DC: The World Bank.

Otis, R.J. (1986). *Small Diameter Gravity Sewers: An Alternate for Unsewered Communities*. Report No. EPA/600/2-86/022. Cincinnati, OH: Environmental Protection Agency.

Otis, R.J. and Sirotiak, K. (1987). Sewer-septic tank hybrid promises savings. *Civil Engineering*, **8**, 74–76.

Simmons, J.D., Newman, J.O., Rose, C.W. and Jones, E.E. (1982). Small diameter, variable-grade gravity sewers for septic tank effluent. In *On-Site Sewage Treatment: Proceedings of the Third National Symposium on Individual and Small Community Sewage Treatment* (The Palmer House, Chicago, 14–15 December 1981), pp. 130–138. St Joseph, MI: American Society of Agricultural Engineers.

Smith, R. and Eilers, R.G. (1970). *Cost to the Consumer for Collection and Treatment of Wastewater*. Water Pollution Control Research Series No. 17090. Washington, DC: Environmental Protection Agency.

South Australia Department of Health (1968). *Common Effluent Drainage Design*. Adelaide: Common Effluent Drainage Section, SADH.

South Australia Health Commission (1986). *Public Health Inspection Guide No. 6: Common Effluent Drainage Schemes*. Adelaide: SAHC.

10

The Colombian ASAS System

José Henrique Rizo-Pombo

10.1 INTRODUCTION

ASAS is the acronym for *Alcantarillado Sin Arrastre de Solidos,*
Spanish for solids-free sewerage. The system consists of compact
septic tanks and sewers. It was developed in 1981 by the author
as a low-cost solution for a large area of the Colombian city of
Cartagena and a neighbouring town, where severe physical and
socio-economic conditions prevailed. The system's development
was part of a rehabilitation programme undertaken by the
Colombian government with the financial aid of the World Bank.

  The ASAS system turned out to be an efficient, reliable and
low-cost system with a long useful life, appropriate for the
sanitation of urban areas with equal or less severe conditions
than those encountered during the Cartagena project. In view of
this, we developed general design criteria to enable the system to
be used in any location, especially where conventional sewerage
is prohibitively expensive.

10.2 DEVELOPMENT OF THE ASAS SYSTEM

10.2.1 Origin of ASAS

Between 1977 and 1986 the Colombian Government, with
technical and financial support from the World Bank and other
agencies, carried out a rehabilitation programme in the Southeast
Zone of the city of Cartagena and in the town of Pasacaballos.
The aim was to improve the physical and social environment and

provide an appropriate level of public services. In that part of the Southeast Zone with less severe physical and socio-economic problems, the programme aimed to provide conventional sewerage. For the rest of the zone and for Pasacaballos, with their greater limitations in terms of physical and socio-economical conditions, the programme sought a more suitable sanitary solution.

The Southeast Zone comprises a large section of Cartagena and lies adjacent to a very shallow lagoon, 2200 ha in area, which is highly polluted by several discharges from the Cartagena sewerage.system. In 1981, the part of the Zone under the rehabilitation programme had 45 000 very poor inhabitants occupying 260 ha of a flat area along 6.5 km of the lagoon shore. The average width of the populated area was 400 m, with elevations slightly above the lagoon water level. Here the soil is soft (0.1–0.2 kg/cm^2 bearing capacity), plastic and consisting of impervious clay and lime combined with refuse material. The water table is very high.

Pasacaballos is located on the mouth of Canal del Dique, a branch of the Magdalena river, on the Bay of Cartagena. In 1981 the population, also very poor, was 4000 and occupied 40 ha. Soil conditions of approximately half of the area are similar to those of the Southeast Zone, with the remainder being better in all aspects.

Finding a sanitation solution appropriate to the conditions of both areas was the object of the project 'Study of Sanitary Solutions in the Southeast Zone of Cartagena and the Jurisdiction of Pasacaballos' which the author's consulting firm carried out between March 1981 and November 1982.

10.2.2 Execution of the study

Our broad approach to the study entailed gathering information on two principal factors: first, which technologies existed; and second what physical, economic, social and cultural conditions prevailed in the areas of interest. With this information, we would be able to select technologies appropriate to the conditions. To ascertain the relevant conditions in the areas of interest we interviewed representatives of the official housing agency and members of the community chosen at random. We also carried out a census of every household in the area, as well as collating detailed information on the social, geographical and demographic environment, including such factors as existing sanitation, projected population increases, typical house characteristics and so forth.

10.2.3 Evaluation of potential solutions

Some 60 different systems were assessed for their potential usefulness in the prevailing conditions of the two areas under study. High technology, expensive solutions, such as vacuum, oil-flushed or chemical-based systems, were immediately deemed inappropriate. Other systems were discarded for a variety of reasons, usually because of their high cost, but also because of health risks, social attitudes or complicated technical operation. Following this process of elimination, there remained conventional sewerage or septic tanks with some means of coping with excess liquids and sullage. Although technically adequate, a conventional sewerage system was ruled out on the grounds of cost, especially given the nature of the landscape: in flat areas the need for pumping stations to compensate for low falls increases construction, operation and maintenance costs. Furthermore, conventional sewers require large volumes of water, also adding to the cost to the consumer. It was decided, therefore, to opt for a system based on septic tanks.

10.2.4 Preliminary design of the ASAS system

We decided to explore the possibility of combining septic tanks with a non-surface system for the evacuation of liquids. This would have to incorporate some sort of unconventional sewer, able to meet the appropriate efficiency criteria while remaining low cost.

A literature search was carried out, but little useful practical information on unconventional sewers was found. Regarding septic tanks, little information was found about internal processes, especially on accumulation, stabilization and compaction of sludge and scum, and its relation to the composition and concentration of solids in house drainage so that a reliable design could be adjusted to the specific conditions of the population. In general terms, the design of tanks based on the recommendations contained in the technical publications we obtained would have been too big and too expensive.

We were also unable to find any information on the characteristics of sewage produced by populations in the socio-economic strata corresponding to those of the Southeast Zone and Pasacaballos. It was not practicable to investigate directly the sewage characteristics in Cartagena, because those areas with the appropriate conditions did not have sewerage. However, records of general characteristics of Cartagena sewage were available.

With no means to define parameters for a realistic design, we decided to devise a set of preliminary parameters based on theoretical assumptions. A more elaborate design was presented in September 1982. For that report the acronym ASAS was then adopted for the system. Initially, the name ASAS was applied only to the sewers. Subsequently, when the detailed hydraulics and functions of the septic tanks and sewers were better understood, it was decided that the name should refer to the whole system.

10.2.5 ASAS pilot projects

Two small pilot schemes were constructed in December 1981, one for six households in the neighbourhood of Fredonia of the Southeast Zone and the other for four households in Pasacaballos. Both included cylindrical and rectangular septic tanks and were put into operation in early January 1982.

Because the houses had no sanitary installations, compact cabins were built with internal pour-flush toilets and showers, and low-cost sinks and laundry troughs on the outside. The precise layout of the units varied according the preference of the householder. In one of the houses, for example, a bathroom was being built with a cistern-flush toilet and a shower compartment; these were therefore incorporated into the project. In all the houses, the drains from the sinks and pour-flush toilets were connected upstream of the septic tank. The showers and the wash troughs were connected downstream of the septic tank, except in one case where both discharged into the septic tank.

10.2.6   Final design

While the pilot projects confirmed that the ASAS concept and the theoretical design were sound, a number of modifications were needed. For example, unforseen quantities of soil entered the system through the showers and laundry troughs, so we decided to connect these drains directly to the septic tank. Furthermore, the presence of odours indicated the need to 'inoculate' or 'seed' the tanks. Dry horse manure gave good results. In addition a ventilation pipe was installed as a precaution. It also became clear that it would be convenient to insert a cleaning access junction or control box with a screen in the septic tank discharge pipeline to the sewers.

Observations on the use of the sanitary units and fixtures, water consumption and operation of tanks and sewers were made. Measurements of sludge and sullage were taken, and analysis of BOD_5, pH and suspended, dissolved and total solids were made on samples of the septic tank effluents and the final sewer effluent. The BOD_5 ranged between 72 and 184 mg/l, with an average of 140 mg/l. The suspended solids results in the tank effluents when only toilets and kitchen sinks were connected to the tanks ranged between 130 and 160 mg/l. At the sewer outlet, when the showers and laundry troughs were included, the readings were very close to 40 mg/l. Compared with the sewage of the city, which had an average of 250 mg/l for both BOD_5 and suspended solids, reductions of 40–60 per cent for BOD_5 and about 84 per cent for suspended solids were obtained. For sludge accumulation rates, the calculated average was 10.2 litres per person per year. Scum formation was erratic. In those tanks where scum was detected, the average was 3.5 litres per person per year.

Based on these results, adjustments were made to the ASAS design for the Southeast Zone and Pasacaballos, and were presented in the final report of the study presented in September 1982.

10.2.7   Subsequent observations and adjustments

During 1983 measurements of sludge and scum were made in the pilot projects. These indicated very little formation of scum and an accumulation rate of sludge of little more than 10 litres per person per year, largely confirming the earlier observations.

When the rehabilitation programme was concluded in 1986, no agency assumed official responsibility for the ASAS pilot projects. During a visit to Pasacaballos in 1988 it was found that the sewers had been removed and the connection boxes had been destroyed. Subsequent housing development in the area resulted in many homes being demolished and many of the householders changing houses, effectively terminating the project. In January 1995, two houses still had their septic tank, but it discharged directly into the street gutter. It was, however, encouraging to note that these two septic tanks have not required desludging and were found to be functioning properly. This was also the case with the other septic tanks used in the project until they were abandoned.

It has become clear that the main functions of the septic tanks are to retain solids and mitigate the sudden impact of toilet discharges; while the stabilization of organic matter is important and is usually attained to an acceptable degree, it is not the main criterion on which to base the size of the tank. For this reason, in the technical specifications relating to the ASAS system, septic tanks are more accurately defined as interceptor tanks.

10.2.8 New projects

Since the pilot projects, the ASAS system has generated considerable interest both in Colombia and abroad. The Colombian Government commissioned a further two pilot projects to investigate the social acceptability of the ASAS technology as an option for small towns and villages and low-income urban areas. Two ASAS systems have been built in Granada and San Zenon (2500 and 1500 people, respectively) located in the northern region, about 200 km south of Cartagena. The systems began operating in May 1995, and the experience during construction and initial operation of both systems has been excellent in terms of social acceptance, community participation and confirmation of the hydraulic design of the system. In addition, simple, low-cost equipment for desludging the septic tanks has been developed. Manuals for operating and maintaining the systems have been prepared, including instructions for monitoring and cleaning the tanks. A training programme for officers of the local public utilities was also carried out.

10.3 SYSTEM DESIGN

Three principal factors govern the design of the ASAS system: (a) a technical solution appropriate to the social and behavioural needs of the population; (b) hydraulic integration of the interceptor tanks and the sewers; and (c) low cost, while being simple to maintain and operate.

10.3.1 Social factors

Urban residential areas and small towns tend to be more homogeneous than heterogeneous in terms of their socio-economic characteristics, cultural background and behaviour. This homogeneity increases as the socio-economic level decreases. Culture is an important factor in determining living habits, water consumption and personal and domestic hygiene. When the socio-economic conditions of a community improve, hygiene practices usually remain almost totally unaltered.

In low-income urban communities households tend to have a similar number of occupants; a typical household, with the average number of occupants, can thus be taken as a representative unit. (The variation in the number of occupants depending on the time of year, which takes place in many households, does not alter the average.) Houses in such communities will have a similar number and type of plumbing fixtures. Typically these will comprise a shower and a pour-flush toilet (although some may have a cistern-flush toilet), a sink and a laundry trough or tray.

10.3.2 Sewer size

Estimation of the loading of the sewerage system is based on the following assumption: in communities of low socio-economic status, it can be assumed that the average number of occupants per household exceeds the number of plumbing fixtures.

Therefore, during peak hours all fixtures will be in use and the greatest possible discharge rate will occur when a toilet is flushed. Like all natural phenomena, the likelihood of several households producing their maximum discharge exactly at the same time is ruled by the law of probabilities. The simultaneity is determined by the intermittence in the use of the toilet. Because the statistical probability of exceeding the design load is low, sewers can be designed to flow full to obtain small diameters.

10.3.3 Criteria for interceptor tanks

As with the sewers, the design of the interceptor tanks depends on social considerations and cost and efficiency criteria. It is more convenient to adopt one size of tank for the unit house, rather than to have several sizes of tank depending on the number of occupants in each household. The size of the unit tank is determined according to the optimum period for de-sludging taking into consideration construction costs (this was 6 years for the Cartagena, Granada and San Zenon projects). As both construction and maintenance costs must be paid by the householders, the lowest cost must be calculated in terms of monthly rates. The monthly rate must cover construction costs and desludging costs.

10.3.4 Minimum cost parameters

The minimum cost of the system corresponds to the smallest possible interceptor tanks and the smallest possible sewer diameter. To determine that cost, iterative calculations are necessary because, while both the tank size and the sewer diameters are directly related to rates of flow and the solids characteristics, the dimensions of the tank themselves influence the rate of flow due to the tank's attenuating action on toilet flush-water flows and the characteristics of unretained solids.

Computations made in the Cartagena study, working with discrete particles of clay or lime (which are the commonly encountered materials in the areas under study) with a specific gravity of 2.65, enabled us to arrive at a set of basic parameters for the minimum cost of tanks and sewers. Further verification of costs in Cartagena and elsewhere, including the pilot projects in Granada and San Zenon, indicated that these parameters remained sound in spite of drastic changes in prices in Colombia since 1982.

1. *Particle size.* The particle size to which minimum cost parameters correspond is 0.02 mm.

2. *Unit load.* The unit load for the sewer design was calculated for cylindrical interceptor tanks, these being cheaper to mass produce. The rate of flow of the unit load is produced by the simultaneous discharge of all the fixtures, that is, the toilet, the kitchen sink, the laundry trough and the shower. Colombian taps and showers discharge about 0.1 l/s each for normal in-house water pressures (8–10 m water). Thus for these three fixtures the compound discharge is 0.3 l/s.

For the unit tank adopted for the pilot project (diameter: 0.90 m; length: 1.48 m; average occupancy per house: 6.8 people) the maximum possible discharge increment was 0.028 l/s for cistern-flush toilets (17 l) and 0.007 l/s for pour-flush toilets (4 l). The compound rates are, therefore, 0.328 and 0.307 l/s, respectively. For rectangular tanks of equivalent capacity to the cylindrical ones, the maximum discharges are less: the tanks have a greater surface area and therefore toilet flush-water flows have less of an effect on changes in the level of the liquid.

Because there is so little variation in maximum discharge in relation to the type of toilet and tank shape, a conservative unit load of 0.328 l/s, rounded to 0.33 l/s, was then adopted as a basic parameter for ASAS design.

10.3.5 Simultaneity equation and design loads

A suitable method for calculating the probability of simultaneous discharges is described by Gallizio (1964) which determines water demand and drainage in large buildings in which the following three parameters are taken into account to reflect the buffering effect of the interceptor tanks: peak duration, interval between repetitive discharges and duration of discharge. Gallizio's equation is:

$$\log_{10} A^{r-1} - \log_{10} B = \log_{10} Cr^n \qquad (10.1)$$

where A = i/t (i and t in minutes)
     B = h/i (h and i in hours)
    Cr^n = number of possible combinations of r units of a
            total of n
     i = interval between successive discharges
     t = duration of the discharge
     h = duration of peak discharge

The equation was adapted to take account of the following considerations. Although the influence of toilet flushings on the maximum rate of discharge from interceptor tanks is small, they determine the actual maximum value, so the interval between successive uses of the toilet was taken as i. The maximum discharge value prevails only for an instant, but to be conservative the approximate time for the tanks to evacuate the volume of one flushing was adopted for t values. Values for h and i were based on observations and surveys. The curve of simultaneity for 2-hour peaks, $i = 10$ min and $t = 1$ min, is shown in Figure 10.1. Sensitivity analysis with different values for t and i, and with h varying between 1 and 2 hours produced very similar values for r/n. This means that the discharge loads and the capacity of the sewers computed by this method will normally not be exceeded.

As can be seen in Figure 10.1, the curve tends to be asymptotic to 23.5 per cent. In fact, with $t = 30$ seconds, it goes closer to 23 per cent and for values of t approaching 1 second the curve tends to drop to 4 per cent. It was considered too risky to use these low simultaneities, even though in reality the duration of maximum discharge from interceptor tanks lasts only a matter of seconds; so the more conservative values shown in Figure 10.1 were adopted for ASAS design.

Although sewers are usually laid above the watertable, infiltration flows are added to the design load with rates of 0.1 l/s per ha for vitrified clay pipe and 0.05 l/s per ha for plastic pipe. Storm drains are not permitted to be connected to tanks or sewers, but 0.5 l/s per ha is also added to design flows. These rates are half those specified in Cartagena for conventional sewers. It would be feasible to reduce the values for infiltration flows even more but, because the risk of unauthorized connections cannot be ruled out, conservative values are recommended.

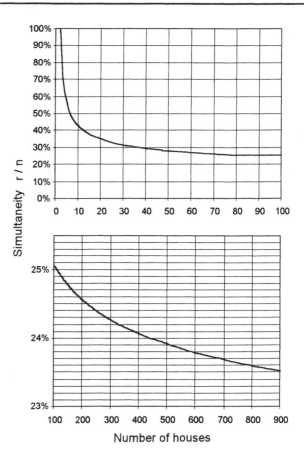

Figure 10.1
Simultaneity curves for
sewer design flows

10.3.6 Minimum velocity and slope

The low content of fine particles in the effluents of interceptor tanks permits velocities as low as 20 cm/s in sewers. Nevertheless, it is not advisable to adopt a slope of less than 0.1 per cent, because of practical installation limitations. This slope allows velocities of more than 20 cm/s for 75 mm diameter and larger pipes flowing full and receiving a unit load of 0.33 l/s. For 50 mm PVC pipes the minimum slope is a little less than 0.2 per cent. As velocities increase when the pipes are not full, the minimum slopes adopted were 0.2 per cent for 50 mm pipes and 0.1 per cent for larger pipes. Tractive tension theory was used to verify that these parameters were suitable for a particle size of 0.02 mm.

10.3.7 Average number of occupants per house

This is the basic parameter for determining the capacity for sludge and scum accumulation in the interceptor tanks. It is calculated by dividing the population of the area by the total number of houses. In most cases the figure should correspond to the most frequent house-to-occupant ratio. The size of the population and number of houses can be established by a census or, more easily but less accurately, by a survey. For Cartagena the average number of occupants per house was 6.8 and for Granada and San Zenon it was 5.76 and 5.78, respectively, rounded up to 5.8.

10.3.8 Location of sewers and interceptor tanks

As shown in Figure 10.2, sewers are typically located on each side of the street in the sidewalk area. House connection boxes, which serve for inspection purposes, are linked by the sewers, so the length is similar to the frontage length of the properties. In narrow streets, sewers can be located more centrally.

Interceptor tanks must be located according to the householders' preference, but must have easy access for inspection and desludging and the shortest possible length for the outlet pipe—preferably without change of direction, as shown in Figure 10.2.

In low-income communities traffic loads are low, so at street crossings the sewers can be laid at a depth of only 10 cm below the gutter, embedded in concrete. The same applies to the entrance of garages. If properly laid in this way, vitrified clay or plastic pipes of up to 100 mm diameter will not be damaged by heavy vehicles. For pipes of larger diameter, the minimum depth is 20 cm. At 50 cm depth, if pipes are properly laid and the fill is well compacted, concrete embedding is not required.

10.3.9 Sewer design

Sewers are designed with the aid of charts as for conventional sewers, also taking into account the number of houses and computing the correct design load using the simultaneity system.

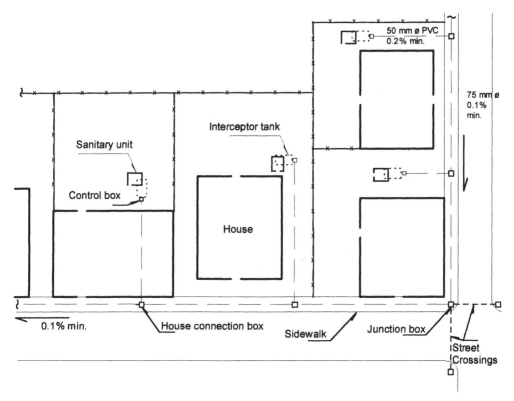

Figure 10.2 Typical ASAS layout

To calculate pipe diameters, any of the standard formulae may be used. That most frequently used in Colombia is Manning's equation.

10.3.10 Interceptor tank design

The dimensions of interceptor tanks are calculated according to Stokes's law for settling tanks. Length, width and minimum liquid depth are proportioned so that the 0.02 mm particle will settle with the maximum discharge rate corresponding to the unit load calculated for the sewer design. This needs iterative calculations because the unit load depends on the surface area of the tank. It is therefore easier to start with a known tank design and adjust its length if the number of fixtures is different. Disturbances caused by vertical drains, even with the sudden discharge

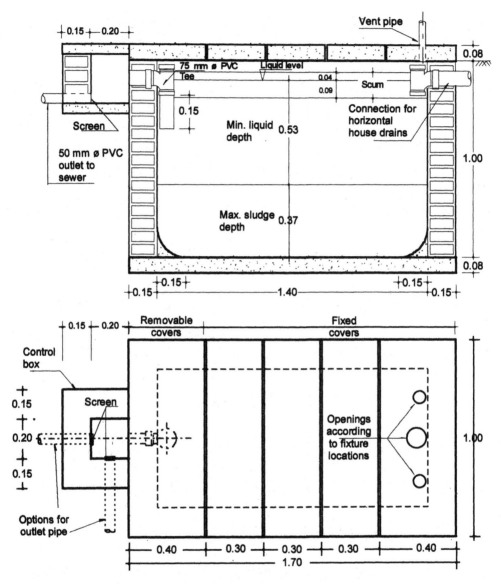

Figure 10.3   Interceptor tank and control box (dimensions in m)

of toilet flush-water flows, can be ignored because the mean velocity of flow in tanks for the maximum discharge is less than 1 cm/s and the surge subsides rapidly. Even if a surge forces some solids out, it is only for an instant and it mixes with discharges from other houses that probably do not carry solids.

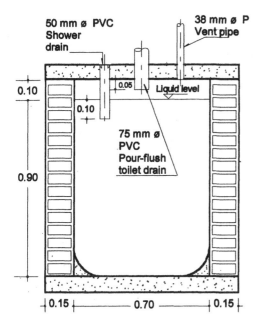

Figure 10.3 (*Continued*)

Tanks can have any shape provided the dimension in the direction of flow permits settling of solids. For projects where mass production is possible, cylindrical tanks may be the cheapest option. If mass production is not possible, rectangular tanks are better and can often be built with the participation of the community. Figure 10.3 shows the tanks built in the Granada and San Zenon pilot project. The interceptor tanks discharge into a small box equipped with a screen on the outlet pipe to retain any floating solids that might escape from the tank (see Figure 10.4).

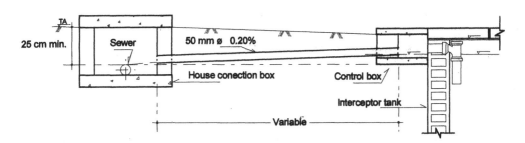

Figure 10.4 Typical house connection

10.3.11 The sanitary unit

The sanitary unit is an important supplementary component of
ASAS projects in low-income communities where houses usually
have no sanitary facilities. The unit (Figure 10.5) is compact but
with sufficient space inside for a pour-flush toilet and a shower
with a floor drain. Outside, a kitchen sink and a laundry trough
or tray back on to the walls. Both are provided with a hose-type
faucet. The location and orientation of the unit and the
positioning of the fixtures are according to the householders'
preference, taking into account easy access from the house and
the location of the interceptor tank. The most economic position
of the unit is over the interceptor tank, so that the fixture drains

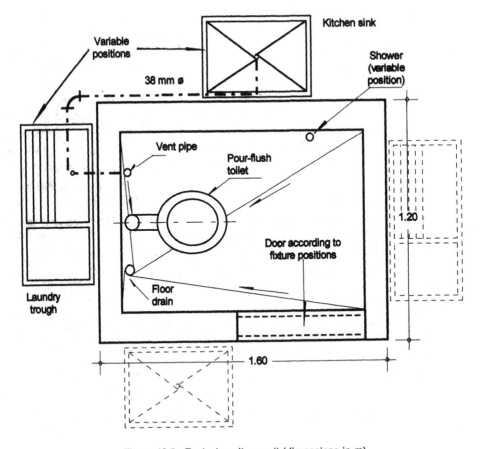

Figure 10.5 Typical sanitary unit (dimensions in m)

discharge vertically into the tank. The pour-flush toilet discharges 5 cm above the liquid level, and the floor drain pipe discharges 10 cm below the liquid level to produce a hydraulic seal (see Figure 10.3). The kitchen sink and the laundry fixture have a common drain connected to the tank vent pipe through a water-seal trap unit.

10.3.12 Maintenance

The maintenance of an ASAS system essentially involves desludging the interceptor tanks. Occasionally, a small piece of cloth or other floating material will need removing from the control box. More rarely, accidental obstruction in a sewer will require attention. Desludging of the interceptor tanks can be done hygienically with vacuum tankers. To avoid the need for large, complex equipment and the need to desludge a large number of tanks at any one time, a desludging programme that enables a few tanks at a time to be cleaned on a daily or weekly basis is recommended. Sludge can be disposed of in a drying bed.

The ratio of the volume of the contents of an interceptor tank to the total daily discharge in an ASAS system is so small (approximately 1–200) that the sludge can be disposed of without problems in a standard treatment facility. For San Zenon, for instance, the ratio is 0.88–192 m^3/day. Even if several tanks are desludged each day, the effect of the added load of solids, BOD_5 and nutrients is negligible.

10.3.13 Wastewater treatment

The interceptor tanks retain about 80 per cent of the solids and reduce BOD_5 by about 40–60 per cent. For this reason the effluents require only simple treatment, for example in facultative waste stabilization ponds that are designed for a lower organic loading than would be the case for conventional sewage.

10.4 ASAS COSTS

For the Cartagena project the ASAS costs (1982 prices) per household were Col\$15 400 (US\$42) for vitrified clay pipe sewers and Col$18 300 (US$50) for PVC, while the cost for conventional sewerage was about Col$36 000 (US$98). In other words the ASAS system was half the cost of conventional sewerage.

In the San Zenon project (1995 prices), the ASAS costs with vitrified clay pipe was Col$262 000 (US$325) and for conventional sewerage about Col$800 000 (US$1000), that is about three times less expensive. The equivalent costs per person are US$56 for the ASAS system and US$172 for conventional sewerage.

The ASAS system does not require additional expense to build house drains if the sanitary unit is built over the intercepting tanks or very close to it. For conventional sewerage the cost per house averages 15–18 per cent more for house drains. If waste-water treatment is included in the cost analysis, the difference is even greater because for ASAS it is less costly.

10.5 CONCLUSIONS

Based on results and observations from the ASAS projects that have been carried out to date, the following conclusions can be made.

1. A minimum slope of 0.1 per cent appears to be adequate and there seems little reason to increase it.

2. A particle size of 0.02 mm has proved an adequate basis for arriving at the minimum cost in all projects. Interceptor tanks which retain that size of particle are sufficiently small (if observed rates of accumulation of sludge and scum are greater than those experienced in the Cartagena project, then the tank dimensions should be enlarged).

3. The number of free-flowing fixtures is the most important determinant for the unit load. For projects with a different number and type of fixtures than described here, this must be taken into account.

The basic parameters described in this chapter have been used successfully in other ASAS projects where the general conditions were similar or less severe than those experienced in Cartagena.

We conclude that ASAS technology is applicable in areas

regardless of socio-economic considerations, and that the numeric values derived for the Cartagena project are applicable in places with similar socio-economic and physical conditions.

10.6 REFERENCE

Gallizio, A. (1964). *Installaciones Sanitarias*. Barcelona: Editorial Cientifico-Médica.

11

Guidelines for the Design of Simplified Sewers

Richard J. Otis, Albert Wright and Alex Bakalian

11.1 INTRODUCTION

Inadequate sanitation is one of the major environmental problems facing urban areas in developing countries today. The unprecedented population growth in urban centres during the last two decades has severely strained the ability of cities to meet the needs for water supply and wastewater treatment. As local governments have tried to cope with insufficient services, priority has been given generally to high-income areas where full or partial cost recovery is considered feasible. Low-income areas are often left unserved or served by woefully inadequate facilities.

The principal reason for this situation is the high cost of conventional sewerage. Treatment alternatives are often considered as a means to reduce total wastewater facility costs, but alternatives to conventional gravity sewers are rarely evaluated, which is suprising as the collection system can represent 65–90 per cent of the total construction costs of collection and treatment. Reducing the cost of sewers would have the greatest impact on the affordability of wastewater collection and treatment facilities. Collection alternatives exist which have been proven to be much less costly but, unfortunately, few engineers are familiar with these technologies.

Development of less costly wastewater collection alternatives have focused on changing the motive force and/or the characteristics of the wastewater collected. Variable grade, small bore effluent sewers, septic tank effluent pumping pressure sewers (STEP) and grinder-pump pressure sewers and vacuum sewers are examples of new alternatives which have been successful

(EPA, 1991). However, modification of conventional design standards have also been used successfully to reduce construction costs substantially. Modifications in design standards have been based on hydraulic theory, improved materials, satisfactory experience, and acceptable risk. 'Flat grade sewers' are one example which have been in use for approximately 80 years in Nebraska (Gidley, 1987). Because of flat terrain and high-groundwater table areas in Nebraska, conventional sewers were expensive. By reducing the minimum slopes or flow velocities, significant cost savings have been realized in construction (shallower sewers and manholes, less dewatering during construction, fewer lift stations) and in operation (lower pumping costs). Another example is 'condominial' sewerage developed in Brazil. Modifications in conventional design standards include reductions in minimum depth, minimum diameter, minimum slope, and changes in service connections. A third example is 'simplified sewerage', which is the subject of this chapter.

11.2 ORIGIN AND DEVELOPMENT OF SIMPLIFIED SEWERAGE

Simplified sewers were developed in Brazil. They are the outcome of changes in several design parameters, including the standards for minimum diameters, minimum slopes, minimum depths, and the spacing and location of manholes. In addition, it makes use of design periods that are considerably shorter than those used in conventional sewerage.

The key impetus for its development was the realization that the application of the conventional design standards was making it difficult to expand coverage to middle- and lower-income communities. As in the USA, the prevailing design criteria were very similar to (and in some cases even more stringent than) those used by Waring in his design of the first separate sewer system in the USA constructed in 1880 (Otis, 1986). The 1880 sewer system, consisting primarily of 150 mm diameter pipes, was designed to carry peak flows at a minimum velocity of 0.6 m/s. Waring argued that, if that velocity was reached at least once a day, the system would remove the 'foul wastes' before they had a chance 'to do mischief' (Waring, 1889). However, to ensure complete removal of deposits, flush tanks were installed at the head of each sewer line. Ventilation was provided through manholes with open grates spaced at a minimum of 300 m apart. Waring's system worked well, the only problems he reported being obstructions caused by objects such as 'a splinter of wood, a carpenter's rule, a bone, a bottle, or some such thing a little longer than the diameter of the sewer'. They occurred primarily in areas near schools and shops. It is interesting to note that most of the original design criteria survived intact (or have become more conservative) in Brazil, with very few exceptions such as the flush tanks and the open grate manholes which have long disappeared. The idea of self-cleaning sewers had become the central design criterion. Unfortunately, the costs of sewer systems based on these century-old criteria have become excessive, prompting engineers in Brazil to question their applicability in the context of their cities.

Consequently, a thorough and critical review of the justification for conventional sewerage design standards was made. This review led to sweeping changes in conventional sewer design standards. The changes were based on a variety of factors such as findings of recent research in hydraulics, satisfactory experience, and redundancy. The outcome of these new standards is a lower-cost sewer system with smaller, flatter, and shallower sewers with fewer and simpler manholes.

11.3 HYDRAULIC DESIGN METHOD

As the major concern in the operation of sanitary sewers is the immediate removal of suspended particles, their design is based on ensuring that a sufficient velocity is constantly maintained to keep the suspended particles moving. In 1940, the Sanitary Section of the Boston Society of Civil Engineers conducted a theoretical analysis of the relationship between bed-load movement and the critical tractive force required to initiate motion of the bed-loads (Boston Society of Civil Engineers, 1942). The committee was able to develop an empirical model for removal of single grain particles of varying specific gravities:

$$V = [(8Kg/f) (S_p - 1) D_g]^{1/2} \qquad (11.1)$$

where V = wastewater velocity, m/s

K = dimensionless parameter with the value 0.4 to initiate motion and 0.8 for adequate cleansing

g = acceleration due to gravity, m/s^2

f = dimensionless friction factor

S_p = specific gravity of the material removed

D_g = particle diameter,

Based on this model, the committee concluded that the self-cleansing velocity is independent of the sewer diameter. As 0.6 m/s was commonly accepted as the necessary velocity which must be achieved to remove grit, the committee supported the practice that flow velocity would be an effective surrogate to the tractive force for calculating slopes of sewers. This approach has been so popular in its simplicity that practising engineers everywhere have used it for decades.

More recent studies by Yao (1974) have shown that there is a direct relationship between the self-cleansing velocity and the critical boundary shear stress or tractive tension:

$$V = (K/n)\ R^{1/6}\ (\tau_c/w)^{1/2} \qquad (11.2)$$

where n = Manning's roughness coefficient

R = hydraulic radius, m

τ_c = critical shear stress, N/m^2

w = specific weight of water, N/m^3

This model indicates that as the diameter of the sewer increases for a given tractive tension, the necessary self-cleansing velocity must increase. Using tractive tensions of 1–2 N/m^2, which appear to be adequate for sanitary sewers, Yao concluded that the practice of using a constant minimum velocity for all sewer sizes results in the underdesigning of larger sewers and the overdesigning smaller sewers.

Using the tractive tension method rather than the minimum velocity method has significant cost implications. For example, if a sewer were designed for initial and final flows of 4.2 l/s and 8.3 l/s respectively, the required slope for a 150 mm diameter sewer is 0.0028 using a tractive tension of 1 N/m^2, as compared with 0.0050 using a minimum velocity of 0.6 m/s. For a trench width of 0.65 m over a length 1000 m, the savings in excavation through the use of the tractive tension would be 1040 m^3. In addition, the downstream end of the trench would be 3.2 m shallower which could provide further savings in rock excavation, dewatering costs or in the elimination of lift stations.

Although the use of the tractive tension method has been advocated for over 20 years, it was not routinely used until the São Paulo, Brazil, water company developed a simple tractive tension design procedure in 1980 (Machado Neto and Tsutiya, 1985; Bakalian *et al.*, 1994). In this procedure, the design slope for a reach of sewer is determined on the basis of the expected initial flow, assuming a tractive tension of 1 N/m² and a minimum flow depth of 0.2 relative to the pipe diameter:

$$I_{min} = 0.0055 Q_i^{-0.47} \qquad (11.3)$$

where I_{min} = minimum sewer slope, m/m
Q_i = initial wastewater flow, l/s

Based on the minimum slope calculated, the diameter of the sewer is determined using the projected final flow and limiting the depth of flow to no more than 75 per cent of the sewer diameter. A hydraulic table (Table 11.1) simplfies this determination by relating the ratio of the depth of flow to the pipe diameter (d/D) to $(Q_f /1000)/I_{min}^{0.5}$ and $V/I^{0.5}$ (Machado Neto and Tsutiya, 1985; Bakalian *et al.*, 1994). The value of $(Q_f /1000)/I_{min}^{0.5}$ or one near this value, is located in this hydraulic design table where d/D does not exceed 0.75. The final velocity (V_f) is computed from the corresponding $V/I^{0.5}$ value in the table. If V_f is greater than 5 m/s, a new diameter is selected in which the depth of flow is somewhat less than 75 per cent of the pipe diameter. Excessive velocities are avoided because of the concern for air entrapment and cavitation at the joints. Manning's equation is used for these computations with a value of *n* of 0.013. A derivation of this method is provided elsewhere (Bakalian *et al.*, 1996).

To illustrate the differences in the resulting designs between the minimum velocity and tractive tension methods, the size and slope of a sewer are determined for an initial flow of 45 l/s and a final flow of 60 l/s assuming *n* = 0.013. Following the minimum velocity method, assuming 0.6 m/s, the selected pipe diameter to carry 60 l/s is 375 mm with a corresponding slope of 0.0016. From a hydraulic elements chart, d/D is 0.73 and the velocity is 0.58 m/s. Initially, the d/D is 0.6 and the corresponding velocity is

Table 11.1 Hydraulic elements of circular pipes based on Manning's equation[a]

	Diameter (m)																	
	0.100		0.150		0.200		0.250		0.300		0.375		0.400		0.450		0.500	
d/D	$V/I^{0.5}$	$Q/I^{0.5}$	$V/I^{0.5}$	$Q/I^{0.5}$	$V/I^{0.5}$	$Q/I^{0.5}$	$V/I^{0.5}$	$Q/I^{0.5}$	$V/I^{0.5}$	$Q/I^{0.5}$	$V/I^{0.5}$	$Q/I^{0.5}$	$V/I^{0.5}$	$Q/I^{0.5}$	$V/I^{0.5}$	$Q/I^{0.5}$	$V/I^{0.5}$	$Q/I^{0.5}$
0.025	1.0733	0.0001	1.4064	0.0002	1.7037	0.0004	1.9770	0.0006	2.2325	0.0011	2.5905	0.0019	2.7044	0.0023	2.9253	0.0031	3.1181	0.0041
0.050	1.6902	0.0002	2.2147	0.0007	2.6829	0.0016	3.1132	0.0029	3.5155	0.0046	4.0793	0.0084	4.2586	0.0100	4.6064	0.0137	4.9416	0.0181
0.075	2.1968	0.0006	2.8785	0.0017	3.4870	0.0037	4.0463	0.0068	4.5692	0.0110	5.3020	0.0200	5.5350	0.0237	5.9871	0.0324	6.4227	0.0430
0.100	2.6392	0.0011	3.4583	0.0032	4.1893	0.0068	4.8612	0.0124	5.4894	0.0202	6.3698	0.0366	6.6498	0.0435	7.1930	0.0595	7.7163	0.0789
0.125	3.0368	0.0017	3.9792	0.0051	4.8204	0.0109	5.5935	0.0198	6.3163	0.0322	7.3293	0.0584	7.6515	0.0694	8.2765	0.0950	8.8787	0.1258
0.150	3.3999	0.0025	4.4550	0.0074	5.3968	0.0159	6.2623	0.0289	7.0716	0.0470	8.2057	0.0852	8.5664	0.1013	9.2661	0.1386	9.9403	0.1836
0.175	3.7350	0.0034	4.8941	0.0102	5.9286	0.0219	6.8795	0.0397	7.7685	0.0645	9.0144	0.1170	9.4107	0.1390	10.1793	0.1903	10.9199	0.2520
0.200	4.0463	0.0045	5.3021	0.0133	6.4229	0.0287	7.4530	0.0521	8.4161	0.0847	9.7659	0.1536	10.1952	0.1824	11.0279	0.2497	11.8303	0.3307
0.225	4.3371	0.0057	5.6830	0.0169	6.8844	0.0364	7.9885	0.0660	9.0208	0.1074	10.4676	0.1947	10.9277	0.2313	11.8203	0.3166	12.6803	0.4193
0.250	4.6095	0.0071	6.0400	0.0209	7.3168	0.0449	8.4902	0.0815	9.5874	0.1325	11.1250	0.2402	11.6141	0.2853	12.5627	0.3906	13.4768	0.5173
0.275	4.8653	0.0085	6.3752	0.0252	7.7228	0.0542	8.9614	0.0983	10.1195	0.1599	11.7425	0.2899	12.2587	0.3443	13.2599	0.4714	14.2247	0.6243
0.300	5.1059	0.0101	6.6904	0.0298	8.1047	0.0642	9.4046	0.1165	10.6199	0.1894	12.3231	0.3434	12.8649	0.4079	13.9156	0.5584	14.928	0.7396
0.325	5.3324	0.0118	6.9872	0.0348	8.4642	0.0749	9.8217	0.1359	11.0909	0.2209	12.8697	0.4006	13.4355	0.4758	14.5328	0.6514	15.5902	0.8627
0.350	5.5456	0.0136	7.2665	0.0401	8.8026	0.0863	10.2144	0.1564	11.5344	0.2543	13.3843	0.4611	13.9727	0.5477	15.1139	0.7498	16.2136	0.9930
0.375	5.7462	0.0155	7.5295	0.0456	9.1212	0.0981	10.5840	0.1780	11.9518	0.2894	13.8686	0.5246	14.4783	0.6232	15.6608	0.8531	16.8003	1.1299
0.400	5.9349	0.0174	7.7768	0.0513	9.4207	0.1105	10.9316	0.2004	12.3443	0.3259	14.3240	0.5909	14.9537	0.7019	16.1751	0.9609	17.3520	1.2726
0.425	6.1122	0.0194	8.0090	0.0573	9.7020	0.1234	11.2580	0.2237	12.7129	0.3638	14.7518	0.6596	15.4003	0.7835	16.6582	1.0726	17.8702	1.4206
0.450	6.2783	0.0215	8.2267	0.0634	9.9657	0.1366	11.5640	0.2477	13.0584	0.4029	15.1528	0.7304	15.8189	0.8676	17.1109	1.1877	18.3559	1.5730
0.475	6.4337	0.0237	8.4302	0.0697	10.2123	0.1502	11.8502	0.2723	13.3816	0.4428	15.5277	0.8029	16.2103	0.9537	17.5343	1.3056	18.8101	1.7292
0.500	6.5784	0.0258	8.6200	0.0762	10.4422	0.1640	12.1169	0.2974	13.6827	0.4836	15.8772	0.8768	16.5751	1.0414	17.9290	1.4257	19.2334	1.8882
0.525	6.7129	0.0280	8.7961	0.0827	10.6555	0.1780	12.3645	0.3228	13.9623	0.5249	16.2016	0.9516	16.9138	1.1304	18.2953	1.5475	19.6264	2.0494
0.550	6.8370	0.0303	8.9588	0.0892	10.8526	0.1921	12.5932	0.3484	14.2206	0.5665	16.5013	1.0271	17.2267	1.2200	18.6337	1.6701	19.9895	2.2119
0.575	6.9510	0.0325	9.1082	0.0958	11.0336	0.2063	12.8031	0.3740	14.4577	0.6082	16.7764	1.1027	17.5139	1.3098	18.9444	1.7931	20.3227	2.3748
0.600	7.0548	0.0347	9.2442	0.1023	11.1983	0.2204	12.9943	0.3996	14.6735	0.6498	17.0269	1.1781	17.7754	1.3994	19.2273	1.9157	20.6262	2.5372
0.625	7.1484	0.0369	9.3668	0.1088	11.3468	0.2344	13.1667	0.4249	14.8682	0.6910	17.2527	1.2528	18.0112	1.4881	19.4823	2.0372	20.8998	2.6981
0.650	7.2316	0.0391	9.4759	0.1152	11.4790	0.2481	13.3200	0.4499	15.0413	0.7316	17.4536	1.3264	18.2209	1.5755	19.7091	2.1569	21.1431	2.8565
0.675	7.3043	0.0412	9.5711	0.1215	11.5944	0.2616	13.4539	0.4743	15.1925	0.7713	17.6291	1.3984	18.4041	1.6610	19.9073	2.2738	21.3557	3.0115
0.700	7.3663	0.0433	9.6523	0.1275	11.6927	0.2747	13.5680	0.4980	15.3214	0.8097	17.7786	1.4681	18.5602	1.7439	20.0762	2.3873	21.5369	3.1618
0.725	7.4171	0.0452	9.7189	0.1334	11.7734	0.2872	13.6616	0.5207	15.4271	0.8467	17.9013	1.5352	18.6883	1.8235	20.2147	2.4964	21.6855	3.3063
0.750	7.4564	0.0471	9.7704	0.1389	11.8357	0.2991	13.7340	0.5424	15.5088	0.8819	17.9961	1.5990	18.7872	1.8993	20.3217	2.6002	21.8003	3.4436
0.775	7.4835	0.0489	9.8059	0.1441	11.8788	0.3103	13.7839	0.5627	15.5652	0.9149	18.0615	1.6589	18.8555	1.9704	20.3956	2.6975	21.8795	3.5725
0.800	7.4976	0.0505	9.8244	0.1489	11.9012	0.3207	13.8099	0.5814	15.5946	0.9454	18.0959	1.7140	18.8911	2.0359	20.4341	2.7872	21.9209	3.6913
0.825	7.4978	0.0520	9.8246	0.1532	11.9014	0.3299	13.8101	0.5982	15.5948	0.9728	18.0959	1.7637	18.8914	2.0949	20.4344	2.8680	21.9212	3.7983
0.850	7.4824	0.0532	9.8045	0.1570	11.8770	0.3380	13.7819	0.6129	15.5629	0.9966	18.0589	1.8069	18.8527	2.1463	20.3926	2.9382	21.8763	3.8914
0.875	7.4496	0.0543	9.7614	0.1601	11.8249	0.3447	13.7214	0.6250	15.4946	1.0162	17.9796	1.8425	18.7700	2.1885	20.3031	2.9961	21.7803	3.9680
0.900	7.3961	0.0551	9.6914	0.1623	11.7401	0.3496	13.6230	0.6339	15.3835	1.0308	17.8507	1.8689	18.6354	2.2199	20.1575	3.0391	21.6241	4.0249
0.925	7.3171	0.0555	9.5879	0.1637	11.6147	0.3525	13.4775	0.6390	15.2192	1.0391	17.6600	1.8840	18.4363	2.2378	19.9422	3.0636	21.3932	4.0574
0.950	7.2032	0.0555	9.4386	0.1637	11.4339	0.3525	13.2676	0.6391	14.9822	1.0392	17.3851	1.8842	18.1493	2.2381	19.6317	3.0639	21.0601	4.0578
0.975	7.0014	0.0549	9.2134	0.1617	11.1611	0.3483	12.9511	0.6315	14.6247	1.0269	16.9703	1.8618	17.7163	2.2115	19.1633	3.0275	20.5576	4.0096
1.000	6.5784	0.0517	8.6200	0.1523	10.4422	0.3280	12.1169	0.5948	13.6827	0.9672	15.8772	1.7536	16.5751	2.0829	17.9290	2.8515	19.2334	3.7765

[a]With Manning's $n = 0.013$. Units are: Q, m³/s; V, m/s; and I, m/m.

Table 11.2 Comparison of design methods[a]

Design method	Pipe diameter (mm)	Slope (m/m)	Initial velocity (m/s)	Final velocity (m/s)
Minimum flow velocity	375	0.0016	0.56	0.58
Brazilian tractive tension	450	0.0009	0.54	0.58

[a]Based on initial flow of 45 l/s and a final flow of 60 l/s.

0.56 m/s. Using the Brazilian procedure for the tractive tension method, I_{min} is 0.0009 and $Q_f/I_{min}^{0.5} = 2.0$. From Table 11.1, a pipe diameter of 450 mm is selected at a d/D of 0.625. The corresponding $V/I^{0.5}$ is 19.48, which translates to $V_f = 0.58$ m/s. Checking the initial conditions, $Q_i/I^{0.5} = 1.5$ which is equivalent to a d/D of about 0.5, corresponding to an initial velocity of 0.54 m/s. These results are compared in Table 11.2.

11.4 DESIGN STANDARDS

11.4.1 Layout

Project areas are typically defined by individual drainage basins. Each basin may have its own collectors and treatment plant. As funds become available, the individual drainage basins may be connected to a common interceptor for conveyance to a regional treatment plant.

As the sewers are usually constructed in existing developed areas, they are commonly installed under sidewalks on both sides of the street, rather than in the street, to minimize the depth of excavation and the cost of pavement restoration. Also, the sewers are extended only to the last connection, rather than to the end of the block, to save pipe and excavation costs.

11.4.2 Design period and flow

Design period

The concept of a design period to project design flows is not used in developed urban areas. Instead, the saturation population for the particular drainage basin is estimated by assuming five residents per dwelling unit. If the saturation population method is not used, a maximum design period of 20 years is recommended. Although the service life of the sewers can be expected to be greater than 20 years, this shorter design period minimizes problems associated with uncertainty of forecasting population growth and water consumption and the high cost of maintaining large sewers with low flows. The shorter design period also avoids large front-end costs, thereby facilitating financing or enhancing the prospects of expanding sewerage service to more areas with a given amount of investment.

Design flow

Design flows are estimated from available water meter readings, per caput estimates and assumed saturation populations. Initial flows are assumed to equal 80 per cent of the metered water supply in the area. Where no metering data are available, a recommended minimum flow in any reach of sewer is 1.5 l/s. Final flows are projected by assuming 120 persons per hectare in urban areas or five persons per dwelling. Per caput flows of 150 – 200 l/day are used. The estimated average wastewater flows are multiplied by peaking factors to obtain an estimate for design. The initial flow is multiplied by 1.5 which represents a minimum day flow. The final flow is multiplied by 1.8 which is the product of the peaking factors for the maximum day (1.2) and the maximum hour (1.5) to obtain an estimate of the instantaneous peak flow. To these estimates, adjustments for potential clear water infiltration are added assuming 0.5 – 1 l/s per km of pipe.

11.4.3 Hydraulic design

Slope computation

The tractive tension method, as described above is used. A tractive tension of 1 N/m^2 is assumed in the design procedure. Depths of flow are maintained between 20 and 75 per cent of the sewer diameter.

Minimum diameter

In Brazil, 100 mm diameter laterals or branch sewers are being used in residential areas for a maximum length of 400 m. These 100 mm diameter pipes are usually located under the unpaved streets of peri-urban communities.

11.4.4 Depth of sewers

In the simplified system, typical minimum sewer depths are 0.65 m below sidewalks, 0.95–1.50 m below residential streets, depending on the distance from the street centreline and amount of traffic, and 2.5 m below heavily travelled streets. Building elevations are not considered in setting the invert elevation of the sewers. If buildings along the mains are too low to enter the sewer by gravity, it is the responsibility of the property owner to find other means of making a connection. In some cases, where topography permits, it may be possible to connect on the other side of the block if easements can be obtained from the neighbouring owners.

11.4.5 Manholes

Manholes constitute an expensive component of a sewer system. In Brazil, engineers have found that conventional manholes are often unnecessarily expensive. Where possible, conventional

Table 11.3 Type and application of manholes in simplified sewerage

Application	Design
Sewer terminus	Terminal cleanout
Long straight sewer	Inspection cleanout
Change in slope/diameter	Buried concrete box
Major junction	Simplified manhole
Minor junction	Inspection cleanout
Minor drop	Inspection cleanout
Major drop	Simplified manhole
Deep sewer (> 3 m)	Conventional manhole

manholes are replaced with 'simplified' manholes, cleanouts, or buried boxes (Table 11.3). Simplified manholes are similar to conventional manholes except they are reduced in size from 1.5 m diameter to 0.6–0.9 m because the need to enter the manholes by maintenance personnel is eliminated due to the shallower depths and to the availability of modern cleaning equipment. They are used only at major junctions. At changes of direction or slope, manholes are replaced by buried boxes or chambers. House connections are adjusted to serve as inspection devices by installing small box under the walkway and connecting it to the sewer with a 45° ell and wye (the jetting hose can be introduced through this box). These changes in the design and use of manholes have been shown to reduce the costs of a sewer system by more than 25 percent.

11.5 COSTS

Simplified sewers have proven to be substantially less costly than conventional sewers. In many places, cost savings ranging from 20 to 50 per cent have been reported. In the State of São Paulo, Brazil, the first projects have shown a reduction of construction costs of 30 per cent but after about 8 years of experience, the reduction is estimated to be closer to 40 per cent. Within the City of São Paulo, cost savings over conventional sewerage are reported to be 35 per cent by SABESP, the water and sewerage utility of the State of São Paulo. SABESP estimates construction costs for small towns (excluding costs of treatment and house connections estimated to be about US$40 and 50, respectively) are US$150–300 per caput for conventional sewerage and

Table 11.4 Construction costs for selected simplified sewerage projects in Brazil[a]

City	São Paulo	Cardosa	Coraodos	Toledo
Total cost	$1 897 000	$48 000	$68 000	$3,762 000
Cost/metre	$76	$13	$8	$21
Cost/caput	$151	$51	$87	$59

[a]Costs in 1988 US$ dollars

US$80–150 per caput for simplified sewage (1988 dollars). Actual costs for selected projects in Brazil appear in Table 11.4.

11.6 OPERATIONAL EXPERIENCE

Since their first implementation in the Brazilian states of São Paulo and Paraná, simplified sewerage has been subsequently applied in Bolivia (Cochabamba and Oruro), Colombia (Bogota and Cartegena) and Cuba (Matanzas). Although no specific data on operational problems are readily available, it is nevertheless known that no significant problems have been reported. In the city of São Paulo, it has been estimated that there are 75 obstructions per 1000 km of sewers each month. This infrequent occurrence of obstruction gives further support to the policy of minimizing the number of manholes. Engineers in SABESP consider that it would be economical to install only a few manholes initially, with the intention of building additional ones if the need arises (i.e. at points of frequent obstructions). Similarly no problems related to excess hydrogen sulphide generation have been reported from field surveys.

11.7 DISCUSSION

The objective of this chapter was to present information on simplified sewerage which provides a new cost-saving approach to the design of sewer systems, based on Brazilian experience. It is based mainly on rational changes in long-standing traditional sewer design standards. The present conventional engineering practice in sewer design was introduced more than a century ago and has undergone relatively few significant changes since. Engineers in Brazil, who more than a decade ago took a serious look at the rationale for the various design criteria, have found ample room for change and simplification without jeopardizing the operational integrity and safety of the system.

Engineering design is not conceived exclusively on the basis of rigid and exact scientific facts but is often based heavily on empirical data supplemented with probabilistic and risk criteria. The factors of safety which have been embedded in many design criteria (design flow, minimum diameter, depth of sewers, etc.)

need not be the same at all times everywhere in all situations. For example, there is no valid basis to apply the same conservative standards in business districts (where breakdowns and repairs could create heavy economic losses and great inconveniences) as in the outskirts of a city (where the impact of similar breakdowns is less severe). In addition to economic aspects, the probability of breakdowns should be a prime consideration in design of a sewerage system. While Gakenheimer and Brando (1983) suggest additional research on uncertainty as it relates to infrastructure standards, they argue that there is enough evidence to move away from the stringent standards that have been adopted from industrialized countries; they contend that 'when resource-limited countries are using conservative standards, risk is lowered in one locality at the cost of fully exposing another'.

With this approach, depending on the prevailing 'engineering culture' and codes, the project engineer still retains the option to apply all or some of the suggested modifications. This review of experience with simplified sewers in Brazil shows that:

- simplified sewerage technology is being successfully applied, and it constitutes a viable lower-cost alternative to the conventional system;

- design modifications that have been introduced in simplified sewerage are based on sound engineering principles and have not compromised the level of service; and

- construction costs have proven to be 30–50 per cent less costly than conventional sewerage allowing sewer service coverage to be extended.

Unfortunately, information on simplified sewerage has not been disseminated much beyond Brazil. It is hoped that in time engineers in other parts of the world will become more familiar with it as increasing operational experience is accumulated and reported.

ACKNOWLEDGEMENTS

The information provided in this paper has been collected, in part, during discussions with the engineering staff of the state water companies of São Paulo (SABESP) and Paraná (SANEPAR).

11.8 REFERENCES

Bakalian, A., Wright, A. and Otis, R.J. (1996). *Sewer Design: The Tractive Force Approach* (in preparation).

Bakalian, A., Wright, A., Otis, R.J. and Azevedo Neto, J. (1994). *Simplified Sewerage: Design Guidelines.* Water and Sanitation Report No. 7. Washington, DC: The World Bank.

Boston Society of Civil Engineers (1942). Minimum velocities for sewers: final report. *Journal of the Boston Society of Civil Engineers,* **29**, 286–299.

EPA (1991). *Alternative Wastewater Collection Systems.* Report No. EPA/625/1-91/024. Washington, DC: Environmental Protection Agency.

Gakenheimer, R. and Brando, C.H.J. (1984). Infrastructure standards. In *Shelter and Development* (ed. L. Rodwin), pp. 133–150. Boston: Allen and Unwin.

Gidley, J.S. (1987). *Case Study No.11: Ericson, Nebraska Flat Grade Sewers.* Morgantown, WV: Small Flows Clearinghouse, West Virginia University.

Machado Neto, J.G. and Tsutiya, M.T. (1985). Tensão trativa: um critério econômico para o dimensionamento das tubulações de esgoto. *Revista DAE,* **45** (140), 73–87.

Otis, R.J. (1986). *Small Diameter Gravity Sewers: An Alternative Wastewater Collection Method for Unsewered Communities.* Report No. EPA/600/S2/86/022. Cincinnati, OH: Environmental Protection Agency.

Waring, G.E., Jr. (1889). *Sewerage and Land-drainage,* 2nd edn. New York: van Nostrand and Co.

Yao, K.M. (1974). Sewer line design based on critical shear stress. *Journal of the Environmental Engineering Division, American Society of Civil Engineers,* **100** (EE2), 507–521.

12

Simplified Sewerage: Simplified Design

D.D. Mara

12.1 INTRODUCTION

Simplified sewerage, often called shallow sewerage, is a low-cost sanitation technology pre-eminently suited to high-density low-income urban areas in developing countries (Sinnatamby, 1986; Sinnatamby *et al.*, 1986; Bakalian *et al.*, 1994). Its comdominial, or backyard, version is less expensive than its in-street version, especially when the sewers of the latter are laid under both sidewalks (Hamer, 1995), although it does require greater agency–customer interaction to ensure good operation and maintenance (Rondon, 1990; Watson, 1995).

In this chapter I wish to show how simplified sewerage theory leads to very simple design, once the basic design parameters are established.

12.2 BASIC THEORY

The minimum tractive tension design approach (Machado Neto and Tsutiya, 1985) will be used as it results in shallower gradients (which may, of course, have greater operation and maintenance requirements), and a fully developed peri-urban area will be considered (i.e. an area in which there is no room for further housing development, so that any future increases in flow are due to increases in water consumption). The two resulting equations are:

$$\tau = \rho g r i \qquad (12.1)$$

where τ = tractive tension, N/m^2 (or Pa)

Table 12.1 Hydraulic elements of a circular section

d/D	k_a	k_r	q/Q
0.02	0.0037	0.0132	0.0007
0.04	0.0105	0.0262	0.0030
0.06	0.0192	0.0389	0.0071
0.08	0.0294	0.0513	0.0130
0.10	0.0409	0.0635	0.0209
0.12	0.0534	0.0755	0.0306
0.14	0.0668	0.0871	0.0421
0.16	0.0811	0.0986	0.0555
0.18	0.0961	0.1097	0.0707
0.20	0.1118	0.1206	0.0876
0.22	0.1281	0.1312	0.1061
0.24	0.1449	0.1416	0.1263
0.26	0.1623	0.1516	0.1480
0.28	0.1800	0.1614	0.1712
0.30	0.1982	0.1709	0.1958
0.32	0.2167	0.1802	0.2217
0.34	0.2355	0.1891	0.2489
0.36	0.2546	0.1978	0.2772
0.38	0.2739	0.2062	0.3066
0.40	0.2934	0.2142	0.3369
0.42	0.3130	0.2220	0.3682
0.44	0.3328	0.2295	0.4002
0.46	0.3527	0.2366	0.4329
0.48	0.3727	0.2435	0.4662

ρ = density of wastewater, kg/m^3
g = acceleration due to gravity, m/s^2
r = hydraulic radius, m
i = sewer gradient, m/m

and:

$$q = k_1 k_2 w N p / 86400 \qquad (12.2)$$

where q = peak flow, l/s
k_1 = peak factor (= daily peak flow/average daily flow)
k_2 = return factor (= sewage generation/water consumption)
w = water consumption, litres per caput per day (lcd)
N = number of households served
p = average household size

Table 12.1   (Continued)

d/D	k_a	k_r	q/Q
0.50	0.3927	0.2500	0.4999
0.52	0.4127	0.2562	0.5340
0.54	0.4327	0.2621	0.5684
0.56	0.4526	0.2676	0.6029
0.58	0.4724	0.2728	0.6374
0.60	0.4920	0.2776	0.6718
0.62	0.5115	0.2821	0.7059
0.64	0.5308	0.2862	0.7396
0.66	0.5499	0.2900	0.7728
0.68	0.5687	0.2933	0.8054
0.70	0.5872	0.2962	0.8371
0.72	0.6054	0.2987	0.8679
0.74	0.6231	0.3008	0.8975
0.76	0.6405	0.3024	0.9257
0.78	0.6573	0.3036	0.9524
0.80	0.6736	0.3042	0.9773
0.82	0.6893	0.3043	1.0003
0.84	0.7043	0.3038	1.0209
0.86	0.7186	0.3026	1.0390
0.88	0.7320	0.3007	1.0541
0.90	0.7445	0.2980	1.0657
0.92	0.7560	0.2944	1.0731
0.94	0.7662	0.2895	1.0755
0.96	0.7749	0.2829	1.0712
0.98	0.7816	0.2735	1.0566
1.00	0.7854	0.2500	1.0000

Now $r = k_r D$, where k_r is a function of the proportional depth d/D (Table 12.1) and D = sewer diameter (m). Thus equation (12.1) can be written as:

$$D = (\tau/\rho g)/k_r i \tag{12.3}$$

Manning's equation is:

$$v = (1/n)r^{2/3}i^{1/2} \tag{12.4}$$

where v = velocity of flow, m/s
       n = Manning's roughness coefficient

As $q = av$ where q = flow (m³/s) and a = cross-sectional area of flow (m²), equation (12.4) can be written as:

$$q = (1/n)ar^{2/3}i^{1/2}$$
(12.5)

Writing $r = k_rD$ and k_aD^2 where k_a is a function of d/D (Table 12.1):

$$q = (1/n)k_aD^2 (k_rD)^{2/3} i^{1/2}$$
(12.6)

Substituting equation (12.3) in equation (12.6):

$$q = (1/n)k_ak_r^{-2} (\tau/\rho g)^{8/3}i^{-13/6}$$
(12.7)

The minimum gradient, I_{min}, is given by rearranging equation (12.7) and writing $i = I_{min}$ and $\tau = \tau_{min}$:

$$I_{min} = [(1/n) k_ak_r^{-2}]^{6/13}[\tau_{min}/\rho g]^{16/13} q^{-6/13}$$
(12.8)

Equation (12.6) can be rearranged, with $i = I_{min}$, as:

$$D = n^{3/8}k_a^{-3/8}k_r-1/4 (q / i_{min}^{1/2})^{3/8}$$
(12.9)

12.3 NUMBER OF HOUSEHOLDS SERVED

Examination of Table 12.1 shows that the proportional flow q/Q (where q is the flow in the sewer at any given depth of flow d and Q is the flow at $d = D$) is given as follows:

for $d/D = 0.6$: $q/Q = 0.6718$

and $d/D = 0.8$: $q/Q = 0.9775$

So designing simplified sewers to flow now with a d/D of 0.6 allows for an increase in water consumption of $[(0.9775-0.6718)/0.6718]$, i.e. 46 per cent, for example from 100 to 146 lcd, which seems a more than adequate allowance.

Equations (12.8) and (12.9) can now be solved for $d/D = 0.6$ (i.e. from Table 12.1, $k_a = 0.4920$ and $k_r = 0.2776$), with $\tau_{min} = 1$ Pa, $n = 0.013$, $\rho = 1000$ kg/m³ and $g = 9.81$ m/s², and expressing q in l/s rather than m³/s:

$$I_{min} = 5.18 \times 10^{-3}q^{-6/13}$$
(12.10)

$$D = 0.0264 \, (q \, / \, I_{min}^{1/2})^{3/8} \qquad (12.11)$$

Substituting equation (12.10) in equation (12.11), rearranging and changing the units of D from m to mm:

$$q = 9.8 \times 10^{-5} \, D^{13/6} \qquad (12.12)$$

Equation (12.2) can be expressed using, as typical values, $k_1 = 1.8$, $k_2 = 0.85$, $w = 100$ lcd and $p = 5$, as follows:

$$q = 0.009N \qquad (12.13)$$

where N = number of households served.

Substituting equation (12.13) into equation (12.12) gives:

$$N = 10.89 \times 10^{-3} \, D^{13/6} \qquad (12.14)$$

12.4 SIMPLIFIED DESIGN

Table 12.2 can be established using equations (12.12) and (12.14), and this can be used directly to design each stretch of sewer, as follows:

• determine the number of households to be sewered;

• select, from Table 12.2, the sewer diameter;

Table 12.2 Maximum number of households served by simplified sewers of 100-300 mm diameter[a]

Sewer diameter (mm)	Maximum number of households served
100	234
150	565
225	1360
300	2536

[a]based on equation (12.14) (see text for inherent values of parameters, such as household size, assumed).

- determine the peak flow from equation (12.13); and

- calculate the minimum sewer gradient from equation (12.10) with $q \nleq 2.2$ l/s (see Sinnatamby, 1986), and ensure that the actual sewer gradient is not less than this.

Table 12.2 is only valid for the choice of parameter values made. For ease of reference, these are:

$$
\begin{aligned}
d/D &= 0.6 \\
k_1 &= 1.8 \\
k_2 &= 0.85 \\
w &= 100 \text{ lcd} \\
p &= 5 \\
\tau_{min} &= 1 \text{ Pa} \\
n &= 0.013 \\
\rho &= 1000 \text{ kg/m}^3 \\
g &= 9.81 \text{ m/s}^3
\end{aligned}
$$

For any other value(s), an equation similar to equation (12.14) has to be derived, and Table 12.2 recalculated for any desired range of sewer size.

12.5 REFERENCES

Bakalian, A., Wright, A., Otis, R. and de Azevedo Netto, J. (1994). *Simplified Sewerage: Design Guidelines*. Water and Sanitation Report No. 7. Washington, DC: The World Bank.

Hamer, J. (1995). *Low Cost Urban Sanitation in Developing Countries*. B.Eng Dissertation. Leeds: University of Leeds (Department of Civil Engineering).

Machado Neto, J.G.O. and Tsutiya, M.T. (1985). Tensão trativa: um critério econômico para o dimensionamento das tubulações de esgoto. *Revista DAE*, **45**, 73–87.

Rondon, E.B.N. (1990). *A Critical Evaluation of Shallow Sewerage Systems: a Case Study in Cuiabá, Brazil*. MSc(Eng) Dissertation. Leeds: University of Leeds (Department of Civil Engineering).

Sinnatamby, G.S. (1986). *The Design of Shallow Sewer Systems*. Nairobi: United Nations Centre for Human Settlements.

Sinnatamby, G.S., McGarry, M.G. and Mara, D.D. (1986). Sewerage: shallow systems offer hope to slums. *World Water*, **9** (1), 39–41.

Watson, G. (1995). *Good Sewers Cheap? Agency-Customer Interactions in Low-Cost Urban Sanitation in Brazil*. Water and Sanitation Currents. Washington, DC: The World Bank.

13

Water Conservation: the Impact of Design, Development and Site Appraisal of a Low-volume Flush Toilet

J.A. Swaffield and R.H.M. Wakelin

13.1 INTRODUCTION

13.1.1 Plumbing and water conservation: then and now

Water supply and sanitation, as a fundamental prerequisite of urban civilization, may be traced to 8000 BC Jericho, progressing via the Indus Valley, Greece, Crete, where 2000 BC Knossos is often taken as a prime example, to Rome where the daily water supply is quoted in the range 700–1300 litres per caput per day (Billington and Roberts, 1982).

Modern water supply and drainage practice may be taken as post-Chadwick, the Industrial Revolution providing both the cause for many of the public health problems associated with the failure of water supply and sanitation provision in the urban centres and the reformist movements that worked to alleviate these problems. The latter part of the nineteenth century saw the final development of the cistern flush toilet and the trap seal, the introduction of vented building drainage systems and the establishment of reliable water supplies to the urban areas of the industrially developed countries. It is interesting to note that many of Chadwick's recommendations were delayed due to the privatized nature of the water supply 'industry' in 1840.

It is perhaps sobering to recall the concerns of the plumbing community at the end of the nineteenth century, as recorded by Hellyer who condemned 'oversized drains, insufficient flows to provide self cleansing, drains that run up hill, or through 90° bends and junctions, lack of trap seal retention, inefficient WC's and the eight minutes needed to refill a cistern' (Wright, 1980).

Water conservation was also an issue at this time, with the London Water Companies attempting to establish the flush volume at 9.1 litres, when up to 40 litres had been in use. The *Builder* journal at the time condemned this as a retrograde step (Billington and Roberts, 1982). This exchange of views parallels recent discussions in the UK concerning a reduction of flush volume from 9.1 to 6.0 litres, which has met an almost identical reaction from the modern ceramic industry (British Bathroom Council, 1988).

WC evaluation tests also date from this period: Robinson (Billington and Roberts 1982) recorded tests in 1893 at the Sanitary Institute where a 9 litre flush left 5 per cent of solids in the trap and 21 per cent along a 50 foot drain, these values reducing to 1 per cent and 5 per cent with a 13.5 litre flush volume. Such tests are now common, with their results being the subject of considerable discussion in the UK, USA and elsewhere.

Water conservation is a necessary and achievable objective. Forty per cent of domestic water in the UK is flushed via the WC. The WC is the one appliance where meaningful savings could be made by a careful synthesis of hydraulic and design methodologies. This chapter describes one such successful water conservation project.

13.1.2   Definition of need and project background

The hydraulic parameters governing low water use WC designs were studied by the authors from 1978 to 1981, during the development of a 4.5–6 litre flush volume close-coupled WC for the UK market, funded by the UK Confederation of British Ceramic Sanitaryware Manufacturers (Uujamhan, 1981). In 1983 the authors initiated a 5-year research project, funded by the UK

Overseas Development Administration (ODA), to design, develop and site-appraise a low-volume flush WC, with the cooperation of Caradon Twyfords Ltd (Bocarro, 1988).

Previous experience and discussions with ODA established the following criteria for the project:

1. A flush volume of 3–4.5 litres would be realistic.

2. A washdown pedestal P-trap WC pan flushed by an independent cistern fitted with a drop valve would offer the widest application. The potential to use locally available cisterns and components was essential to ensure availability of spare parts in the future.

3. The potential to offer a bucket/pour flush WC pan, upgradable to cistern flush, should be considered. The omission of a flushing rim, which could not be cleaned by pour-flushing, would then be necessary and an alternative method of flush water distribution developed (in fact this pour-flush development was little used and further discussion of it is omitted in this chapter).

4. The necessity for simplicity of design and operation to ensure adequate maintenance of the WC. This ruled out systems such as vacuum- or pressure-assisted solids transportation, siphon break-valves and diaphragms.

5. The predominance of single storey housing and buildings in developing countries minimizes pressure fluctuations in the drainage system and would allow reduced trap-seal depth, subject to evaporative loss constraints.

6. Gaborone in Botswana and Maseru in Lesotho would be used for site evaluation of the low-volume flush toilet (LVFT) under a wide range of conditions.

This chapter presents the research and development necessary to produce an appliance within these guidelines, together with an overview of its site evaluation in Botswana, where the installation conditions were closest to those in the UK.

13.2 LABORATORY DEVELOPMENT OF THE LOW-VOLUME FLUSH TOILET

A P-trap WC configuration was adopted, to conform to common practice in developing countries. The modelling of the LVFT was based on a Twyfords British Standard 1213 P-trap WC from which both the flushing rim and trap-seal back plate had been removed. Plasticine was used to remodel the internal shape of the bowl and trap, and perspex back plates were fitted to vary trap-seal depth. A mechanical tipping bucket was developed to provide repeatable and representative pour-flushing by bucket. In addition to British Standard 5503 tests, which included the ball test, paper test and sawdust test, a range of discriminatory tests were used and developed to appraise WC performance.

1. Multiple ball test, using fifty 20 mm diameter balls of the same specific gravity which could be selected in the range 0.85–1.15. Discharge efficiency was calculated as the percentage of balls discharging successfully.

2. Liquid contamination test, using potassium permanganate to simulate liquid contamination of the WC trap. Light absorption, measured by a colorimeter, of a sample of trap-water taken after flushing compared with a sample taken before flushing, and calculated as a percentage relative residual concentration, gives an important measure of flushing performance.

3. Modified paper test, using six pieces of 125 × 125 mm newspaper.

4. Simulated faeces test, using three 30 mm diameter by 100 mm long foamed plastic stools having a saturated specific gravity of 0.98.

5. Modified ball test, using the BS5503 ball but recording the amount of water discharged ahead of the ball relative to the flush volume. This modification produced a useful degree of discrimination.

6. Various bowl washing tests, including the American ANSI A112.19.2M (1982) test, which involves drawing a horizontal

line round the bowl 25 mm below the rim with a water-soluble ink pen and measuring the total length of unwashed line after flushing.

7. Blotting paper test, to measure splashing from the bowl during a flush by placing a sheet of blotting paper over the WC bowl and either estimating the area of wetted paper or weighing the paper before and after flushing.

8. Transport test, using a 14.3 m length of 110 mm UPVC discharge pipe at a gradient of 1 in 80, connected to the WC. Transportation performance of a model solid, typically a half-length C2-type maternity pad, was monitored by recording its velocity profile along the pipe (Howarth *et al.*, 1980). This test provides the basis for the definition of the required flush volume relative to length and gradient of drain prior to connection of discharge from other sanitary appliances.

With the omission of the flushing rim, it was necessary to develop an effective means of distributing the flush water around the bowl. A device termed a 'diverter bar' was developed with two slightly downward angled side slots, with the primary function of distributing cleansing water round the bowl, and a bottom jet to induce momentum transfer to solids in the trap. The diverter bar was connected via a sleeve to allow for different diameters of flush pipe.

The design criteria governing low-volume flush WC performance (η) are shown in Figure 13.1. The four parameters, trap-seal volume (S), flush volume (F), trap-seal depth (h) and minimum trap passage clearance (w), can be functionally related as follows:

$$\eta = f\,(Sh/Fw) \qquad (13.1)$$

Uujamhan (1981) identified two dimensionless parameters, n_S and n_F, determining WC performance for solid or fluid contamination removal, respectively (Figure 13.2), which should be as low as possible.

Laboratory testing and development led to a final WC prototype termed the Mark III WC and a Mark VI diverter bar, finally appraised with a plastic cistern and a modified ceramic

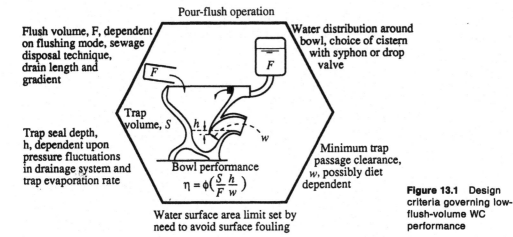

Pour-flush operation

Flush volume, F, dependent on flushing mode, sewage disposal technique, drain length and gradient

Water distribution around bowl, choice of cistern with syphon or drop valve

Trap volume, S

Trap seal depth, h, dependent upon pressure fluctuations in drainage system and trap evaporation rate

Bowl performance

$$\eta = \phi\left(\frac{S}{F}\frac{h}{w}\right)$$

Minimum trap passage clearance, w, possibly diet dependent

Water surface area limit set by need to avoid surface fouling

Figure 13.1 Design criteria governing low-flush-volume WC performance

Vaal Potteries cistern, produced in South Africa. The ceramic cistern was adopted for the site trials due to the fire risk caused by candles placed on a plastic cistern and also as no indigenous manufacture of sanitary fittings exists in Botswana and Lesotho. The plastic and ceramic cistern fitted bowls (termed the Mark III and 'field' Mark III WC, respectively) were compared with the original Twyfords BS1213 WC (termed Mark 0) for flush volumes between 3 and 9 litres, as shown in Figures 13.3 and 13.4.

Figure 13.3 illustrates the 0.85 specific gravity multiple ball test. The Mark III WCs performed similarly, and even at 3 litres

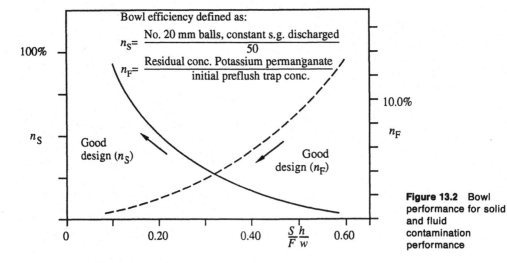

Bowl efficiency defined as:

$$n_S = \frac{\text{No. 20 mm balls, constant s.g. discharged}}{50}$$

$$n_F = \frac{\text{Residual conc. Potassium permanganate}}{\text{initial preflush trap conc.}}$$

100%

10.0%

n_S

n_F

Good design (n_S)

Good design (n_F)

0   0.20   0.40 $\frac{S}{F}\frac{h}{w}$ 0.60

Figure 13.2 Bowl performance for solid and fluid contamination performance

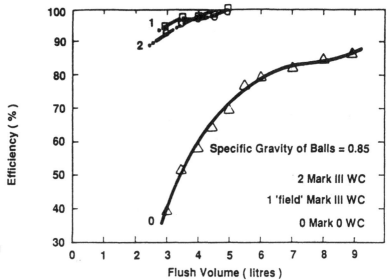

Figure 13.3 A comparison between the multiple ball efficiencies obtained by the Mark 0, Mark III and 'field' Mark III WCs

performed better than the original (Mark 0) Twyfords WC flushing with 9 litres. Figure 13.4 presents the results of the liquid contamination test, with a lower residual concentration for the Mark III WCs at 3 litres than the Mark 0 WC at 9 litres. The Mark III WCs only required 2 litres to pass the BS5503 paper and ball tests, but required 3.5 litres to pass the BS5503 sawdust test.

Results from the solid transportation tests are shown in Figure 13.5, presented as solid velocity along the drain, set at a gradient G at distance L from the WC drain connection, plotted as $\sqrt{(L/G)}$, providing a linear characteristic (Swaffield and Wakelin, 1976). Figure 13.5 indicates that a flush volume of 3 litres should only be adopted for drain lengths up to 11.25 m at a gradient of 1 in 80 prior to inflow connection from other sanitary appliances. Most on-site sewage disposal or storage systems would meet this restriction. For drains connected to a main sewer, a minimum flush volume of 4 litres was recommended to achieve similar solid transportation performance to the Mark 0 WC flushed with 9 litres. These flush volumes were adopted for the site trials.

These results confirmed Uujamhan's (1981) functional relationship (equation 13.1 and Figure 13.2) for solid and fluid contamination removal. From Figure 13.2 the value of the dimensionless

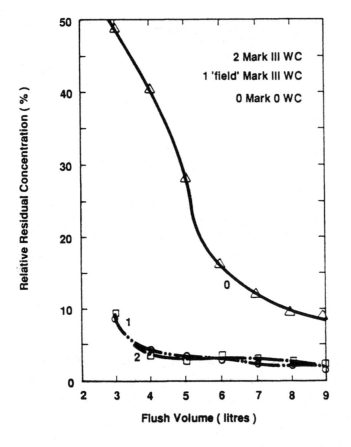

Figure 13.4 A comparison between the concentration test results of the Mark 0, Mark III and 'field' Mark III WCs

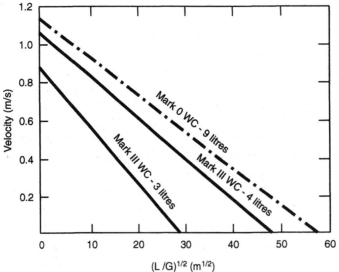

Figure 13.5 A comparison between the solid transport results of the Mark 0 and Mark III WCs

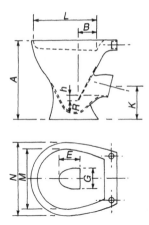

Description	BS 5503: pt 2	Twyfords BS 1213 p - trap WC pan Mark 0	Mark III WC
Depth of water seal *h*	50 min.	56	36
Clearance below tip of plate *R*	75 min.	77	69
Trap seal volume	n/a	1.79 L	0.86 L
Water surface:			
back to front *E*	150 min.	165	118
side to side *G*	110 min.	113	79
Height *A*	390 + 10	406	390
Distance from tip of back plate to inside face of flush rim *B*	70 max.	30	32
Width of opening *M*	240 max.	260	275
Length of opening *L*	290 max.	320	320
Width of opening *N*	360 + 10	355	350
Height to centre of WC outlet *K*	190	191	190

Figure 13.6 A comparison between the prototype WC and British Standard designs. All length measurements are in millimetres

parameter should be as low as possible, subject to certain physical limitations. It was decided to adopt a minimum trap passage width (w) of 63 mm in order to meet dietary constraints. Consideration of drain pressure fluctuation and trap-seal evaporative loss suggested that trap-seal depth (h) could be safely reduced from 50 to 36 mm. These two limitations dictated a minimum trap volume (S) of approximately 0.86 litres. The value of the dimensionless parameter (Sh/Fw) for the Mark 0 WC with 9.1 litre flush is 0.143 (Figure 13.6). Substituting the limits suggested above would require a flush volume (F) of 3.4 litres to attain similar WC performance, a prediction confirmed by the tests mentioned above.

13.3   SITE TRIALS IN BOTSWANA

Caradon Twyfords (Stoke on Trent, UK) produced 220 of the Mark III WC pans (Figure 13.6), and 220 Mark VI diverter bars

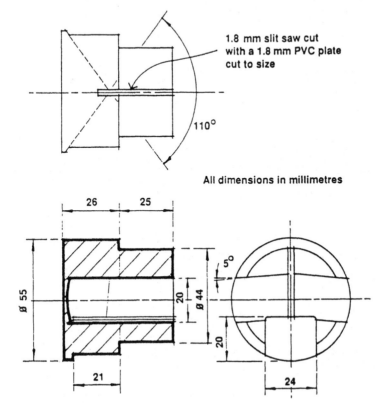

1.8 mm slit saw cut with a 1.8 mm PVC plate cut to size

110°

All dimensions in millimetres

Figure 13.7 The Mark VI PVC diverter bar

(Figure 13.7) were machined from PVC. The overseas site trials, which commenced in August 1985, involved ceramic cisterns and were concentrated in the capitals of Botswana and Lesotho. The Botswana programme of prototype WC installations was completed by May 1986. The main concentration of installations was in Gaborone where some 95 of the units were installed.

There were several aspects of the monitoring. Monthly inspections of all the project installations were undertaken in Gaborone to identify any problems with the WCs. These inspections solved a problem with the cistern design caused by a lack of quality control in manufacture.

Three drainage surveys were carried out on an estate of 63 houses fitted with the experimental WCs in Gaborone West (Figure 13.8). The surveys showed that a WC flush volume of 4 litres had no adverse effect on any of the main sewer lines or the house-to-main-sewer connections, which were to British Standards.

Water consumption was monitored on this estate. Water readings were taken at 14-day intervals from December 1986 to

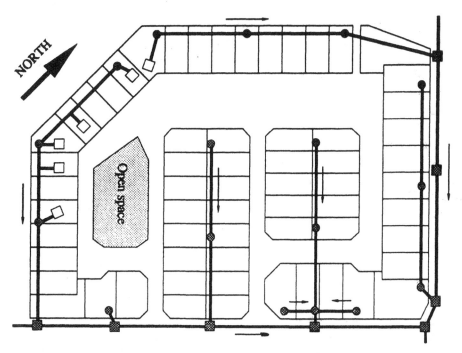

Figure 13.8 Layout of 63 low income Botswana Housing Corporation house plots, Gaborone West, fitted with 4 litre flush WCs

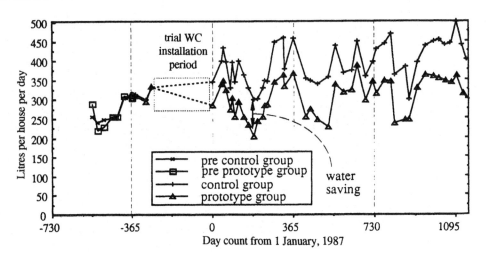

Figure 13.9 Comparison of prototype 4-litre flush w.c. housing group's water consumption relative to a control group's over a 3 year period. Also note pre-installation water consumption is sensibly identical for both groups

September 1987 and thereafter at 1 month intervals to January 1990. An adjacent control sample of 30 identical houses with 10-litre WCs was also monitored. The prototype group were found to use a mean of 2.4 kilolitres less per house per month than the control group, or a difference of 19 litres per caput per day. In terms of percentage savings, the prototype houses used 16–30 per cent less water than the control group during December 1986 to January 1990. Figure 13.9 shows consistent seasonal water savings and also illustrates that the water consumption for the prototype and control groups prior to the installation of the trial WCs was sensibly identical over the 12-month period from February 1985.

Additional water data collected to determine the percentage of household water that was used for WC flushing indicated that the prototype group used 21 per cent and the control sample used 40 per cent for this purpose. This result may be used to predict the savings possible for a range of flush volumes. From a knowledge of the total water use during any period in both prototype and control group dwellings, and the appropriate percentage used for WC flushing, the non-WC usage may be determined. The number of WC operations for the prototype WC is also known from its flush volume, thus the hypothetical

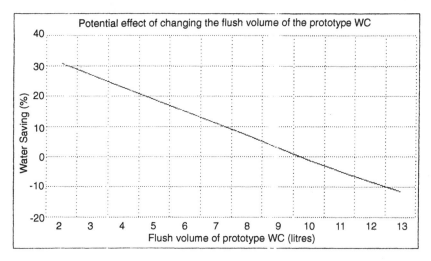

Figure 13.10 Potential effect of changing the flush volume of the prototype WC

WC water use with any flush volume may be added to the non-WC volume used. Figure 13.10 demonstrates this graphically, although caution must be exercised in order to avoid reducing the flush volume to the point at which double WC flushing becomes necessary. A study of the number of WC operations inherent in the control and prototype groups via flush volume setting and WC total usage illustrates that the prototype WC was normally only flushed once per usage, indicating the efficiency of the design. From a detailed questionnaire designed primarily to assess user reaction to the LVFT, it was apparent that the two groups were virtually identical in terms of population structure, age, employment, income and education, justifying the selection of the two samples for comparison. The overall reaction to the prototype WCs was positive.

13.4 CONCLUSIONS

Water conservation via the use of LVFT is an achievable objective utilizing known design methodologies and materials. User reaction, particularly where the introduction of LVFT is linked to cash savings in reduced water costs, was shown to be favourable.

The most important result of the 1200-day monitoring exercise in Gaborone West was the total lack of drainage system blockages arising from the use of the LVFT, thus destroying, for a system built to British Standards, the myth that water conservation automatically leads to blocked drains.

The LVFT design installed in Gaborone and Maseru (Lesotho) has also now been supplied to China, where the Building Research Institutes of Shanghai and Chendu are responsible for trial installations, and to Brazil, where a 30-house site is being monitored by the Instituto de Pesquisa Tecnológica, the technology research institute of the State of São Paulo, as part of a collaborative research study with Heriot-Watt University.

13.5 REFERENCES

Billington, N.S. and Roberts, B.M. (1982). *Building Services Engineering: A Review of its Development*. London: Pergamon.

Bocarro, R.A. (1988). *Water Conserving WC Design for Developing Countries*. PhD thesis. Uxbridge: Brunel University.

British Bathroom Council (1988). Press release relating to introduction of 6 litre w.c. flush volume, reported in *The Sunday Times*, 'Endangered: The Great British Loo', 24 November.

Howarth, G., Swaffield, J.A. and Wakelin, R.H.M. (1980). Development of a flushability criterion for sanitary products. In *Drainage and Water Supply for Buildings* (CIB/W62 Symposium) (ed. R.H.M. Wakelin). Uxbridge: Brunel University (Drainage Research Group, Department of Building Technology).

Swaffield, J.A. and Wakelin, R.H.M. (1976). Observation and analysis of the parameters affecting the transport of waste solids in internal drainage systems. *The Public Health Engineer*, 4(6), 165–170.

Uujamhan, E.J.S. (1981). *Water Conservation WC Design: a Study of the Design Parameters Affecting WC Performance*. PhD thesis. Uxbridge: Brunel University.

Wright, L. (1980). *Clean and Decent: The History of the Bath and the Loo*. London: Routledge and Kegan Paul.

14

Third World Surface Water Drainage: The Effect of Solids on Performance

P.J. Kolsky, J.N. Parkinson and D. Butler

14.1 INTRODUCTION

Effective sanitation of any kind requires stormwater drainage. Pit latrines, septic tanks and sewers can transmit disease when flooded, as their contents mix with runoff and overflow into the streets and homes of the community. In recognition of these and other problems with urban flooding, the Engineering Division of the British Overseas Development Administration (ODA) funded a 2-year project entitled 'Performance-based Evaluation of Surface Water Drainage in Developing Countries'. This chapter summarizes some of the most salient findings from this research in the city of Indore in India.

14.1.1 Significance of flooding

Before looking at the technical details of the work, flooding and its consequences should be put in context. How serious are the risks to public health from flooding? Which aspects of flooding most trouble residents?

Water quality data

Stormwater quality is one sign of the risks from surface water flooding. As background data for this study, 57 grab samples of flood water were taken from a total of 19 locations in two catchments in Indore during four storms in August and September 1994. These were analysed for faecal coliforms (FC) using the MPN technique (Indian Standards Institution, 1982). All samples showed FC concentrations greater than 10^5/100 ml, 55 were greater than 10^6/100 ml, and in the heaviest flood, half the 18 samples were above 10^7/100 ml. These samples were not taken from combined drains themselves, but from the streets of two communities where flooding took place.

Water quality standards for bathing water and for agricultural reuse of wastewater provide a useful point of comparison, as they involve comparable types of human contact. While adults do not deliberately swim in minor floods, exposure still occurs during children's play and adult travel in the community. In addition, flood waters frequently enter lower-lying homes, spreading faecal contamination into the domestic environment. Besides the relatively short-lived bacterial contamination during flooding, it is also plausible that flooding contributes to *Ascaris* and other enteric worm infections through the spread of their eggs.

Until recently, European bathing water standards were set at 2000 FC/100 ml (Council of the European Communities, 1976); these have recently been tightened (although not without controversy) to 200 FC/100 ml. Water quality guidelines for agricultural reuse are set at 1000 FC/100 ml, largely to protect the health of consumers, but also to remain consistent with bathing water quality standards (Mara and Cairncross, 1989). By either standard, the concentrations measured in the flood waters of Indore catchments thus violated both bathing and reuse standards by three to four orders of magnitude.

Community perceptions

In another part of the study, Stephens *et al.* (1994) used qualitative social science methods to investigate perceptions of flooding by members of four Indore communities. Some of their findings are of particular interest to engineers:

> Flooding was ranked low in comparison to other risks and problems, such as improvements in job opportunities, provision of housing, mosquitoes and smelly backlanes.

A major concern mentioned by residents of all four areas relates to the predictability of the flooding event. ... In other words, even extensive inundation is bearable if expected. ... Interventions aimed at ameliorating the effects of flooding should try to take account of these needs of the community to understand and adapt their coping strategies if necessary.

Residents in flat areas give equal, if not more, weight to the after-effects of rains. ... as compared to immediate effects. Water may stand for long periods in very flat areas. Respondents feel that this makes walking difficult and allows the breeding of mosquitoes. Faeces-contaminated mud caused by stormwater is seen as most problematic as it is also a perceived source of mosquitoes and noxious smells.

[In areas of the slum improvement programme...] residents had high expectations of drainage improvement when the projects were initiated and appear to have expected that flooding and inundations would cease or be reduced substantially. It is likely that the feeling that their expectations have not been met is in part an immediate sense of disappointment that flooding has not ceased altogether.

[Drainage] interventions would gain favour if residents understood and were clearly informed about the effects (good and bad) on the environmental risk which they perceive as inherently a natural event. This, of course, necessitates technical personnel being able to predict the consequences of technical interventions. Such a strategy might reduce the scale of expectation.

14.1.2 Conventional drainage practice

While the extensive literature of urban stormwater management describes many sophisticated techniques of analysis (e.g. DoE/NWC, 1982; WEF/ASCE, 1992; Marsalek and Torno, 1993), drainage design as practised for small catchments is quite straightforward. Conventional practice defines the problem as designing a system to pass a flow generated by a storm of a specified return period or frequency. With this approach, the engineer can only hope to say 'This system's capacity will be exceeded on average once every x years' (but even x can be difficult to estimate with available data). The engineer cannot rationally predict how high the water will rise, how big an area will be flooded, or how long a flood will last. If residents are to understand that drainage networks cannot eliminate flooding, engineers need to know and explain what is likely to happen when it does flood.

Understanding performance during floods matters less where flooding is rare. Flooding in slums of developing countries, however, is a frequent occurrence for a variety of reasons. Rainfall is more intense in tropical climates than in temperate ones (Watkins and Fiddes, 1984; Shaw, 1994), but the funds available for drainage improvement projects are scarcer. In addition, the capacity of most slum drainage networks is greatly diminished by heavy solids deposits. These solids often include sediment loads from unpaved areas, construction debris, road metal and solid waste. While many have noted this problem, (e.g. Cotton and Franceys, 1991; Kolsky *et al.*, 1992; Cairncross and Feachem, 1993) there has been no previous systematic study exploring the impact of such solids upon the performance of the drainage system as experienced by the community.

14.1.3   The research questions of this study

The notion of performance

What are appropriate measures of drainage system performance? As noted above, the current standard measure is the return period of the design storm, which is inadequate for prediction of what will happen when it floods. Equally seriously, most engineers do not try to minimize the effects of such flooding when it inevitably occurs. Pioneering work has been done to develop the 'major/minor' or dual drainage design philosophy (Wisner and Kassem,1982; Argue, 1986) in which the role of (major) surface flow during flooding is explicitly recognized. However, field application of these ideas to slum drainage in developing countries has been minimal. Our initial definition of drainage performance as the depth, area, duration and frequency of flooding evolved in light of what could be measured and modelled.

The role of solids

Rubbish, construction debris and sediment inevitably end up in drains. What is their impact upon flooding? If drainage performance can be defined and measured, it may serve as a basis on which to evaluate the impact of solids. Given hydrological uncertainties, this cannot be done on an experimental basis; no two storms are identical, and experimental control over the level of solids in a real drainage system is limited. A computer model of the drainage system could, however, allow the user to do 'hypothetical experiments' or sensitivity analyses of performance with varying levels of solids.

Research questions

In summary, the research questions addressed in this chapter are:

- Can the performance of a conventional open storm drainage system in Indore be defined and modelled?

- Can the level and size of solids in this system be usefully characterized?

- Can sensitivity analyses with the model show how performance varies with the level of solids?

- What are the implications for practice of the results from the above?

14.2 MODEL CONSTRUCTION, CALIBRATION AND VERIFICATION

14.2.1 The site

Indore

Indore is a city of 1.2 million people located 125 km southwest of Bhopal in the large central Indian state of Madhya Pradesh. The city was chosen as the study site for several reasons. Both the ODA and the Indore Development Authority expressed initial interest in the work and provided substantial support throughout. The ODA-funded Slum Improvement Programme there applied innovative drainage designs, and it was hoped that the research could compare these sites with conventional practice. A final factor was the strong collaboration offered by the Shri

Govindra Seksaria Institute of Technology and Science. Over the
two monsoons, detailed studies were undertaken on three catchments: two were included in the Indore Slum Improvement
Programme, incorporating innovative design, while the third
reflected conventional local drainage practice. This chapter
describes work done in the conventional catchment, Pardeshipura.

The study catchment

Pardeshipura is a flat, medium density, heterogeneous development of residential and business properties in the northeast of
the city. Within Pardeshipura, a 17-ha study area was selected
which drains to a single major outlet. The study area (Figure
14.1) contains about 5000 inhabitants, who are classed within the
middle-income group, although pockets of the area may be classified as slums. Runoff in the catchment drains to a 585 m long
concrete-lined open channel, which also carries sewage. The
outlet discharges freely under most conditions, simplifying
baseline hydraulic measurement and analysis.

The main drain's nearly rectangular cross-section varies from
approximately 0.9 m deep and 0.6 m wide at the outlet to
approximately 0.5 m deep and 0.4 m wide at the upstream end.
The overall slope is approximately 0.003, but varies irregularly
along the channel. Single or parallel pipes carry flow beneath
intersections, and often act as partially blocked obstructions.
While one major tributary parallel to the main drain was also
modelled, the other tributary drains in the catchment are small,
irregular and hydraulically inefficient. These secondary tributaries
were observed to reach maximum capacity rapidly during even
the smallest storm events, resulting in flooding of the adjacent
streets. The overflow in these streets was therefore modelled as
flow in wide, shallow, V-shaped channels.

Exact boundaries of the catchment were difficult to determine.
The large flat area of the catchment was one problem, as small
variations in topography have a disproportionate influence on
tributary area. Secondly, the tributary area actually varied during
heavy storms when the drainage capacity of adjacent catchments
was exceeded. Under these circumstances, runoff would overflow
into the study area across tributary boundaries. By the same

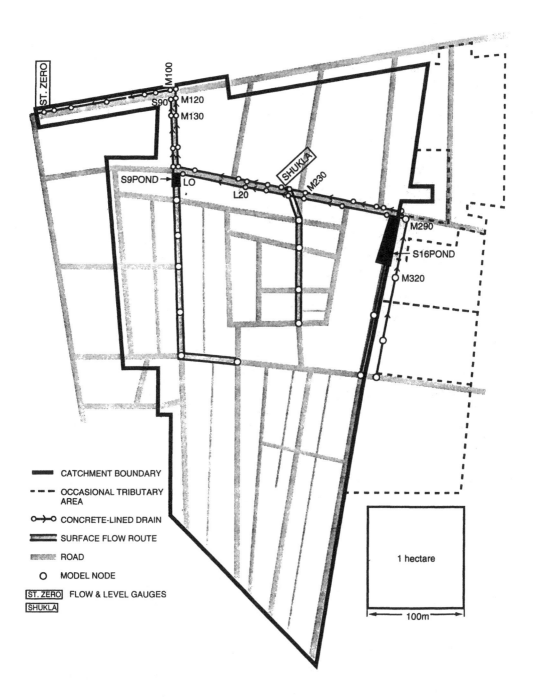

Figure 14.1 Pardeshipura catchment area

token, the capacity of the main drain of the study area was itself exceeded during floods, and substantial amounts of runoff would then cross the area boundary into a neighbouring catchment. This phenomenon was clearly observed on several occasions, and was incorporated in the model.

14.2.2 Selection of hydraulic model

The dynamic drainage simulation model SPIDA (developed by Wallingford Software) was chosen to model the Pardeshipura catchment. SPIDA was chosen because it has some capacity to model flooding, and it has a suitably simple way to incorporate the effects of solids deposition. In addition, SPIDA is a mature hydraulic model that has been widely used and tested, reducing the risks of programming errors inherent in building new software.

Modelling of solids

With SPIDA, the user specifies the depth of solids at each section of the channel. The model then treats this depth as a fixed physical obstruction, with a user-specified sediment roughness. Hydraulic computations thus reflect both the reduced hydraulic cross-section and the increased hydraulic roughness resulting from the deposition of solids in each section. SPIDA assumes that solids levels are fixed throughout the storm, and no attempt is made to simulate sediment transport, erosion, or deposition.

Modelling of flooding

SPIDA calculates a solution to a set of one-dimensional partial differential equations relating discharge and water level in the links and nodes of the modelled network. The model thus computes flows and levels throughout the simulated network at each time step of the simulation. In any given time step, flooding occurs at a node when the water level is predicted to rise above the node's specified ground level.

In reality, the relationship between water level, flooded surface area and volume of flood water is a complex one determined by details of topography. Of necessity, the model simplifies this complexity. At many nodes, the Pardeshipura model assumed SPIDA's simple 'double cone' storage volume, in which the user specifies the surface area of flooding at each of two flood depths. Between the ground level and the lower depth, the storage volume is represented by an inverted cone with its apex at the node centre; the base of this inverted cone has an area equal to the surface area of flooding at the lower flood depth. Above this lower flood depth, the incremental volume is defined by the frustum of a second concentric inverted cone, the angle of which is defined by the slope between the two user-specified depths and areas. At these 'cone' nodes, all flood water is stored, and then returned to the channel at the same node as the water level recedes. Field observation confirmed that flood waters did recede back into the channel in many areas.

In other parts of the catchment, however, most of the water flooding out of the drain was seen to flow across the ground and return to the drain further downstream. For these cases, a hypothetical 'overflow weir' was assumed between the two nodes at the upstream and downstream ends of the overland flow. Finally, at a few downstream nodes in the catchment (M100, M120 and S90), any water leaving the channel effectively leaves the system, as the channel here is the boundary of the catchment. This effect was included in our model.

14.2.3 Data on solids

Solids levels in modelled drains were checked twice weekly, with periodic spot checks to find out if rapid variations took place during the week. These measurements were part of a programme to record solids buildup since a cleaning in September 1993. Solids depths measured in the open drainage channels in the Pardeshipura catchment were high. On average, approximately 30 per cent of the channel depth was filled with solids. However, in some locations, especially towards the upstream end, solids depths were as high as 85 per cent.

Towards the end of the study period, samples were taken from five evenly spaced stations. At each station, a 1 m length was cleaned out, and the contents analysed for their size distribution.

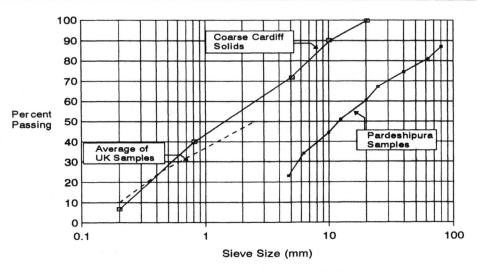

Figure 14.2 Comparison of drain solids size distributions

The averaged results of this analysis are shown in Figure 14.2, in comparison with results from averages of UK sewer sediment studies, and a curve describing a coarse sediment sample from Cardiff (Binnie *et al.*, 1987). The Indore curve is significantly to the right of the UK closed conduit sewer sediment data. This shows that the drain solids in Indore contain more large material; compare the Indore d_{50} of 12.5 mm with the d_{50} of 2.43 mm found as a mean of UK sediment studies (Binnie *et al.*, 1987). Indeed, more than 10 per cent by mass of the Indore drain solids are larger than 80 mm. The application of sediment transport theory to sewer design is still in its infancy (Ackers *et al.*, 1994). It is, however, very likely that material of the size described here will require higher self-cleansing velocities than can be attained in typical open drains.

14.2.4 Hydrological measurement and modelling

Rainfall and runoff measurement

Rainfall was measured by a Montec Detectronics recording rain-gauge (0.2 mm tipping bucket size) at 2 minute intervals. Data were collected between 25 July and 15 September during the monsoon of 1994; during this gauging period, a total of nearly

630 mm of rain was recorded. Rainfall patterns were compared with two other gauges within a kilometre's radius; these patterns appeared consistent, and no significant events appear to have been missed. To measure runoff and hydraulic performance, combined recording level and velocity gauges (Montec DETEC 3510 Surveyloggers) were installed at the outlet and approximately halfway along the main drain. This equipment recorded levels using pressure transducers, and velocities using ultrasound.

Surface run-off modelling

A version of the US Soil Conservation Service TR-55 model (SCS, 1975) incorporated in the SPIDA package was used to model rainfall–runoff relationships. While the more sophisticated 'Wallingford Procedure' also built into SPIDA is more suitable in many applications, it is based on field research in UK catchments under hydrological conditions distinctly different from those encountered in India. The modified SCS method is simpler and easier to calibrate than more sophisticated models, and is also more widely known outside the UK. Two parameters are particularly important in the method for calculating a runoff hydrograph from rainfall over a surface: storage and routing coefficients. For each type of cover, users must specify an 'S' or storage value to compute a 'curve number' which determines the proportion of rainfall converted to runoff. Based on field surveys and observation, four types of cover were identified, and the fraction of the catchment made up by each was estimated. Appropriate 'S' values were estimated during early stages of model calibration, by comparison of predicted and observed

Table 14.1 Types of cover and SCS curve numbers

Type of cover	Percentage of catchment	SCS storage depth (mm)	SCS curve number
Roof areas	34%	6	98
Paved area	27%	8	97
Open ground	24%	40	76
Muddy ground	15%	80	86

runoff volumes and outflow hydrographs (Table 14.1). Routing coefficients for different tributary areas varied with the directness of runoff flow to the main drain, as more distant areas were assigned higher routing coefficients than those draining directly to the channel.

14.2.5 Model calibration and verification

Definition of storm events

The SPIDA model was calibrated and subsequently verified using rainfall, flow and level data collected as described above. Storm events were identified as those having a maximum rainfall intensity average over a 30 minute period of at least 4 mm/h. (A review of level and flow data showed that rainfall of intensities less than this resulted in no significant hydraulic response in the drains.) Using the above definition, 21 storm events were recorded; these accounted for 86 per cent of the total recorded rainfall. These events were subsequently used for both calibration and verification of the model, and for computer simulation to test the sensitivity of performance to different solids levels.

Initial calibration using smaller storms

Seven storms known not to cause major flooding were used for initial calibration. The objective of this work was to tune the hydrological variables of catchment response without the complications of flooding. Predicted and observed level and flow hydrographs were compared at the sites of the two recording level and flow gauges, and adjustments were made to various parameters of the model. Close correlation between observed and predicted volumes and hydrographs showed that the hydrological coefficients assigned to each portion of the catchment were sufficiently accurate.

Second stage calibration with flood events

Modification of the model to reflect performance during floods only started after satisfactory agreement had been achieved with the smaller storms. Shapes of the flood cones were estimated from a review of the area's topography. As described above, overland flow between nodes during floods was simulated by using weir links. Predictions of hydraulic response during floods were sensitive to small changes in these weir levels.

Final verification

Four storms were used for final verification of the model. No adjustment to any model parameter was made during these verification runs, and these storms had not been used during any of the preceding calibration process. Agreement between the predicted and observed response was reasonable. Unfortunately, none of these events were flooding events. Figure 14.3 illustrates the fit of depth and discharge in a calibration and verification storm at the outlet gauge; the fit at the midway gauge was similar.

14.3   THE EFFECT OF SOLIDS DEPOSITION ON DRAINAGE PERFORMANCE

Once a satisfactory verified model had been developed, completing sensitivity analyses on the effects of solids deposition upon performance was relatively simple. Solids depths were specified for channel sections as fixed percentages of channel height, and the model was then used to predict performance. Performance was thus examined with solids levels set at 20, 50 and 80 per cent of channel height. A more challenging aspect of the work, however, was converting the original broad notions of performance into terms that could be effectively modelled and measured over a distributed system. In this section, findings concerning the sensitivity of depth, duration, extent and frequency of flooding are presented. An additional measure of the effect of solids upon performance, the fraction of runoff lost from the catchment as flood water, is also presented.

14.3.1   Depth and duration of flooding

Many nodes flooded during large events. To observe variations in flooding patterns, six nodes were selected for detailed analysis of

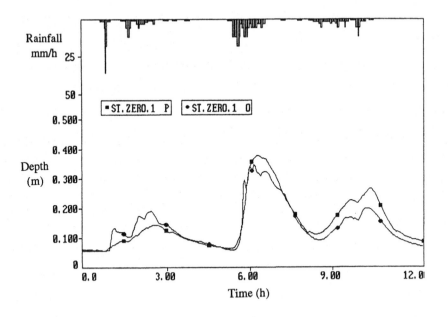

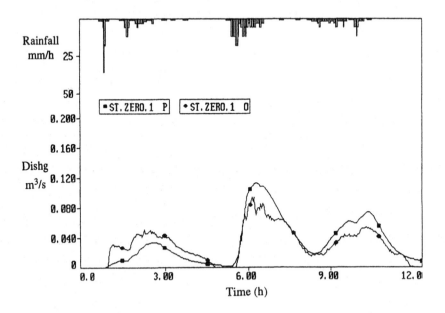

CALIBRATION EVENT: 10/08/94

Figure 14.3   Calibration and verification plots at station 0

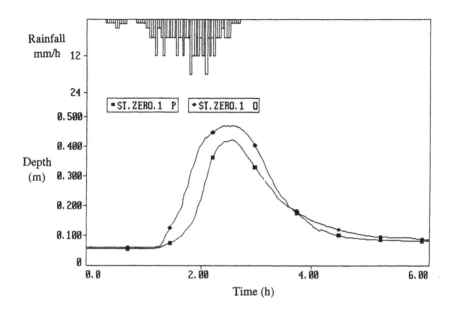

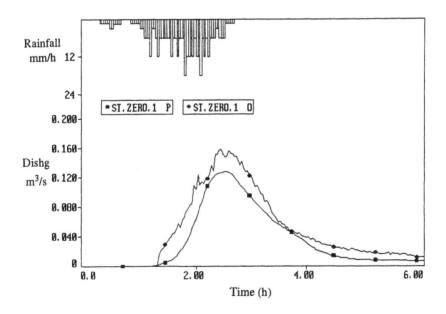

VERIFICATION EVENT: 26/08/94

depth and duration of flooding. 'Flood depth' was defined as the predicted maximum depth of flooding above ground level, while 'duration of flooding' was defined as the period over which water was present above ground level. Both were estimated for the selected nodes using the depth hydrographs produced by the model.

Depth of flooding

Figure 14.4 shows the variation in flooding depth at selected nodes with assumed system-wide solids depths during the largest event. The results show that maximum depth of flooding is, on an absolute scale, insensitive to solids levels. The maximum variation in flood depth between a clean drain and one that is 80 per cent blocked in this large event is only 9 cm. This may be explained by the large area over which water is spread once flooding occurs. In other words, when flooding takes place, the hydraulic capacity of the channel plays a relatively minor role in determining the depth of flooding.

The distributed nature of the drainage system explains the unlikely result of *decreased* depth of flooding as solids increase at

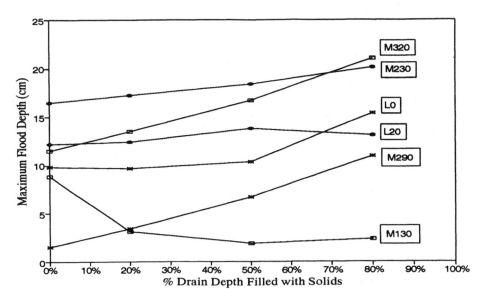

Figure 14.4   Maximum flood depth as a function of solids blockage (event of 5 September 1994)

M130. When the upstream channel is clean, its capacity is maximized and upstream flooding is minimized; this means that the peak flow arrives at M130 with a minimum of attenuation, and thus overflows. When solids partially block the upstream channel, upstream flooding occurs, with a resultant shaving of the peak flow at this node.

In smaller events, channel hydraulics is more important in determining the depth of flooding, but flood depths are also correspondingly lower. The maximum flood depth predicted by SPIDA in all analyses was 0.24 m. It should be noted, however, that 'small increases in depth' can have a disproportionate impact on residents if they overtop the thresholds of their homes!

Duration of flooding

Figure 14.5 shows the sensitivity of flood duration to solids level at these same six nodes. The effect of solids deposition on duration is substantial. In the extreme case, only 2 hours of flooding are predicted at node M230 when the channel is clean, while 15 are predicted when the channel is 80 per cent blocked. Other nodes show less extreme, but still meaningful, variations in

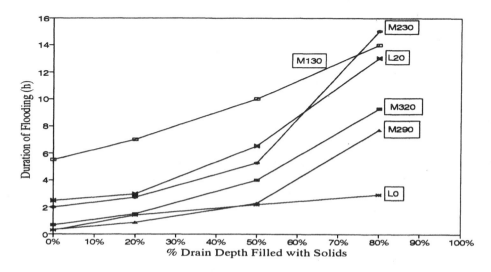

Figure 14.5   Flood duration as a function of solids blockage (event of 5 September 1994)

duration in both relative and absolute terms. For all of these nodes, predicted flood waters return to the channel at some point before leaving the catchment; under these circumstances, it is not surprising that flood duration is sensitive to the capacity of the drainage channel.

14.3.2 Extent and frequency of flooding

Defining extent

Predicting and interpreting the areal extent of flooding from model output is difficult. Extremely detailed topography is required to predict the surface area of flooding, particularly in an urban environment where small changes in depth may lead to step changes in flooded area. In addition, SPIDA's simple flooding models have limited capacity to manage such detailed information, even if it were available.

We adopted an alternative approach of selecting a sample of nodes at uniform intervals along the drainage channel; the extent of flooding can then be gauged by the number of these nodes at which flooding is predicted. The option of counting the total number of flooded nodes was rejected, because some nodes are closely bunched and would therefore effectively count the same body of flood water twice. At the same time, we wished to have a reasonably large sample of nodes that flooded at least once, so that the sensitivity to various levels of solids could be gauged. On this basis, a stationing interval of 55 m was selected, which led to a sample of 15 of a total of 38 channel nodes.

Defining frequency

Development of meaningful intensity–duration–frequency (IDF) curves for Indore rainfall is problematic for a variety of reasons, including restricted access to continuous rain-gauge data. The regional approach of Kothyari and Garde (1992) was explored, but found to predict rainfall intensities far greater than credible in light of local experience. A 'design storm' approach based on conventional return period analysis was therefore unsuitable, and we adopted a more pragmatic approach using the actual rainfall data collected during the study. Frequency of flooding was therefore expressed as a proportion of the total number of storm events observed. This unfortunately cannot be extrapolated to a conventional 'return period'.

A second problem arises in defining independent flood events. Rainfall and water level patterns are not discrete, and distinguishing one flood event from another is sometimes difficult. Suppose, for example, that a high-intensity storm event leads to flooding. After the rainfall intensity recedes, the water level begins a gradual drop. If a second rainfall event of much lower intensity now occurs, a flood may result only because the channel is still nearly full from the antecedent event. For these situations, it was decided that the hydraulic response of the antecedent event was 'over' when the water level had dropped by 90 per cent from its maximum. In two cases, this criterion distinguished two distinct events occurring close together; on the other hand, four storms that lasted 10 hours or more contained multiple, but hydraulically linked, peaks. Because the hydraulic response to the latter peaks could not be distinguished from that of the antecedent rainfall, these longer storms were treated as single events.

The effect of solids

Figure 14.6 shows how the frequency of flooding (as a fraction of recorded storms) and the number of flooded sample nodes increases with increasing solids levels. The graph naturally shows that some nodes are more prone to flooding than others. For any given number of flooded nodes, an 80 per cent blockage of the channel increases the frequency of flooding by a factor between 2 and 3, while a 50 per cent blockage generally accounts for an approximate 50 per cent increase in flooding frequency. Note that this figure does not show anything about the severity (depth and duration) of flooding at any of the nodes, but only that the water rose above ground level.

14.3.3 Fraction of runoff lost from the catchment

Figure 14.7 shows the percentage of runoff lost from the catchment through overland flow as a function of solids blockage

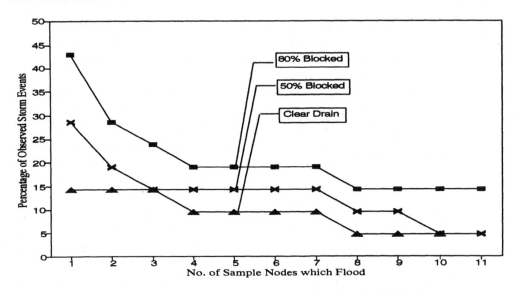

Figure 14.6 Flood frequency and extent as a function of solids blockage

for four events. This does *not* reflect the total volume of flood water, as a substantial volume of flood water stays within the catchment and is then drained by the channel. All of the runoff lost to the catchment, has, however, gone on to flood elsewhere. This figure dramatically summarizes in one graph how solids can limit the capacity of a channel to drain its catchment.

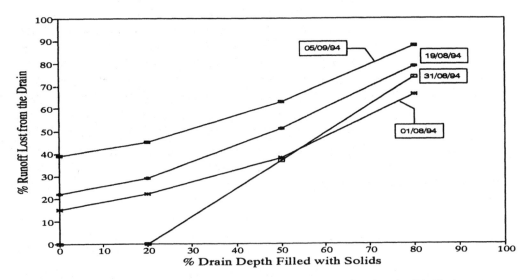

Figure 14.7 Fraction of runoff lost from the catchment as a function of solids blockage

14.4 CONCLUSIONS AND IMPLICATIONS FOR DRAINAGE PRACTICE

14.4.1 Principal findings and their limitations

Flood depth and solids levels

The result that flood depth varies little with solids levels is not surprising. The onset of flooding suddenly increases the 'wetted surface area' over which flow is distributed. If one considers a small drain in a large street, the hydraulic capacity and storage of the drain clearly play relatively small roles in fixing the depth of flow, *once flooding takes place*. As noted in the literature of major/minor drainage systems, (Wisner and Kassem, 1982; Argue, 1986; WEF/ASCE, 1992) the function of the minor drainage system is to reduce the frequency of small floods, rather than to diminish the impact of rare floods.

Flood duration and solids levels

The sensitivity of duration to solids levels depends greatly upon the relative role of surface flow and channel flow in draining flood waters. The proportions of surface and channel flow will vary greatly between catchments, and between different sites within the same catchment. For the nodes examined in this analysis, all flood waters had to return to the channel at some point before leaving the catchment, so that channel capacity has a strong influence on the duration of flooding. At nodes served by a true major/minor system, the additional capacity of the surface drainage route could dramatically reduce the durations observed here. This is a powerful argument in favour of explicit recognition of surface flow in design and analysis; if urban areas are *only* to be drained by conduits, then flood durations will indeed depend upon the channel capacity and will be very sensitive to the degree of its blockage with solids.

Flood extent and frequency and solids levels

The 'extent of flooding' is difficult to model, and, in fact, difficult to measure in the field during a rapidly changing event. Our findings on this aspect of performance are also difficult to interpret, and are most likely to vary from catchment to catchment depending upon topography and the layout of the drainage network. The sensitivity of frequency of minor flooding to solids level, however, is quite clear; this makes intuitive sense given the large number of small storms that can be managed by clean drainage networks of conventional design.

Limitations of the analysis

There are of course questions about both the validity of the findings presented above and their applicability in other contexts. For example, the data are collected from only part of one monsoon, which was generally viewed locally as a 'good' or wet one. How representative is this of Indore, or of other parts of the world? The model also has a limited capacity to simulate the complexity of drainage in Pardeshipura, especially considering the high levels of solids in the drain, and the crude simplicity of the flooding model. Nevertheless, the above findings about performance make sense on an intuitive and physical level, and several had been postulated at the outset of the study (Kolsky, 1992). The overall conclusion is that solids levels significantly affect most measures of performance, and do not affect depth only because other flow routes are available. A question thus arises about how to respond to the almost inevitable capacity reduction of conventional slum drainage networks with deposition of solids.

14.4.2 Inadequacy of current practice

Engineers design drains as if they were free of solids. It is sometimes argued that poor design and construction practice are responsible for the solids deposition evident in so many drains in developing countries; if conventional practice was properly followed, it is argued, such drains would be self-cleansing. We suspect that the solids in drains in developing countries are, like those we measured in Pardeshipura, larger than those in the industrialized world. Construction debris, sediment from unpaved areas, temporary road surfacing materials and solid waste are more likely to enter drains in low-income communities in the developing world than in the industrialized world, where more resources are available to manage these wastes. We also suspect that the notion of 'self-cleansing' velocities is unrealistic for the particle size distributions likely to enter surface water drains in slums of developing countries.

Ignoring solids would not matter if they were limited in depth or extent, or if they had no effect upon drainage performance. Our research and experience strongly suggest otherwise; solids deposition is common, substantial, and greatly affects drainage performance. Surface water drains are *designed* as sumps into which water can flow, but unfortunately serve as sumps for anything else on the ground. While open channels are particularly susceptible, gully pots of closed conduit systems do not stop all solids from entering the drain, particularly if they are not emptied (Butler and Karunaratne, 1995).

14.4.3 An alternative approach

Our research suggests that blockage of drainage conduits is likely to be a recurrent problem, and that frequent surface water flow along the streets of slums in developing countries is practically inevitable. If this is true, engineers should consider the *deliberate* routing of surface water over the streets of the slum, rather than depending on separate pipes or channels to carry these flows. Such an approach has been developed and used by the Indian consulting engineer Himanshu Parikh, who designed the infrastructure for the sites of the ODA-funded Slum Improvement Programme in Indore. Parikh (1990) adopted a variant of the major/minor drainage concept he had developed earlier, and used both pipes and the roads of the site to drain surface water. Given that water will flow over road surfaces anyway, why not try to design them to serve as effective drainage conduits? This may involve more expensive site grading, but will assure more reliable site and road drainage than any system based on conduits that are subject to solids deposition. Roads as drains are also easier to maintain; street sweeping is easier than drain cleaning, and residents have an interest in keeping roads reasonably clear for access. Solids in drains are not an immediate nuisance to residents and are difficult to clean; road obstructions, however, are recognized as more of a problem, and are also easier to remove.

What are the principal limits of such an approach? Clearly, the topography of certain depressions may not permit surface routing of all flows. The approach is also impractical where combined sewerage is, (either by design or illegal practice), the common means of excreta disposal. The choice of road surfacing materials needs to be carefully considered. On sites where engineers are planning low-cost sewerage systems, however, it makes clear sense to minimize the length of drainage channel to be built and maintained, and to design roads to shed runoff away from the catchment as efficiently as possible. For many areas, effective low-cost surface water drainage may therefore best be summarized as 'drainage without drains'.

ACKNOWLEDGEMENTS

This research was funded by the Engineering Division of the ODA, of the UK. However, the ODA can accept no responsibility for any information provided or views expressed. The authors would particularly like to acknowledge Mr Brian Jackson, the ODA Water Resources Adviser, and Messrs David Crapper, Mike Slingsby, and Brian Baxendale of the Slum Improvement Programme Field Management Office in New Delhi offered real encouragement and practical support. We wish to acknowledge the Director and staff of the Shri Govindra Seksaria Institute of Technology and Science for their collaboration in the fieldwork. We are particularly grateful to Professor T.A. Sihorwala, our principal collaborator, to Mr Mansoor Ali, our senior research assistant, and to the staff of the Indore Drainage Evaluation Project who did much of the fieldwork. We also acknowledge the support of the Indore Development Authority, who greatly facilitated the fieldwork. We thank Wallingford Software Ltd, and particularly Dr Roland Price, for both general support and the loan of a copy of SPIDA for this research. Finally, the first author gratefully acknowledges the strong support of Ursula Blumenthal during critical stages of the work, and the general support of Sandy Cairncross throughout, both of whom are in the Environmental Health Group of the London School of Hygiene and Tropical Medicine.

14.5 REFERENCES

Ackers, J.C., Butler, D. and May, R.W.P. (1994). *Design of Sewers to Control Sediment Problems.* Funders Report No. CP/27. London: Construction Industry Research and Information Association.

Argue, J.R. (1986). *Storm Drainage Design in Small Urban Catchments: A Handbook for Australian Practice.* Special Report SR 34. Vermont South: Australian Road Research Board.

Binnie and Partners and HR Wallingford (1987). *Sediment Movement in Combined Sewerage and Storm-Water Drainage Systems*, Project Report No. 1. London: Construction Industry Research and Information Association.

Butler, D. and Karunaratne, S.H.P.G. (1995). The suspended solids trap efficiency of the roadside gully pot. *Water Research*, **29**(2), 719–729.

Cairncross, S. and Feachem, R. (1993). *Environmental Health Engineering in the Tropics: An Introductory Text*, 2nd edn. Chichester: John Wiley.

Cotton, A.P. and Franceys, R. (1991). *Services for Shelter.* Liverpool: Liverpool University Press.

Council of the European Communities (1976). Council Directive 76/160/EEC of 8 December 1975 concerning the quality of bathing water. *Official Journal of the European Communities*, L31/1-7 (5 February).

DoE/NWC (1983). *Design and Analysis of Urban Storm Drainage: The Wallingford Procedure. (Volume 1: Principles, Methods and Practice.)* Department of the Environment/National Water Council Standing Technical Committee Report No. 28. Wallingford: Hydraulics Research.

Indian Standards Institution (1982). *Methods of Sampling and Microbiological Examination of Water*, 1st revision. New Delhi: Indian Standards Institution.

Kolsky, P.J. (1992). *The Impact of Maintenance Upon Drainage Performance in a Low-income Community in India.* PhD proposal. London: London School of Hygiene & Tropical Medicine.

Kolsky, P.J., Hirano, A.P., and Bjerre, J. (1992). *Directions in Drainage for the 1990's.* New Delhi: UNDP/World Bank/WHO.

Kothyari, U.C. and Garde, R.J. (1992). Rainfall intensity–duration–frequency formula for India. *Journal of Hydraulic Engineering, American Society of Civil Engineers*, **118**(2), 323–336.

Mara, D.D. and Cairncross, S. (1989). *Guidelines for the Safe Use of Wastewater and Excreta in Agriculture and Aquaculture.* Geneva: World Health Organization.

Marsalek, J. and Torno, H., (1993). *Proceedings of the Sixth International Conference on Urban Storm Drainage.* Victoria, BC: Seapoint Publishing.

Parikh, H. (1990). *Aranya—An Approach to Settlement Design; Planning* and Design of Low-Cost Housing Project at Indore, India*. New Delhi: Housing and Urban Development Corporation.

SCS (1975). *Urban Hydrology for Small Watersheds*. Technical Release No. 55. Washington, DC: US Soil Conservation Service.

Shaw, E. (1994). *Hydrology in Practice*, 3rd edn. London: Chapman & Hall.

Stephens, C., Pathnaik, R., and Lewin, S. (1994). *This is My Beautiful Home: Risk Perceptions towards Flooding and Environment in Low Income Communities*. London: London School of Hygiene & Tropical Medicine.

Watkins, L.H. and Fiddes, D. (1984). *Highway and Urban Hydrology in the Tropics*. London: Pentech Press.

WEF/ASCE (1992). *Design and Construction of Urban Stormwater Management Systems*. Washington, DC: Water and Environment Federation/American Society of Civil Engineers.

Wisner, P.E. and Kassem, A.M. (1982). Analysis of dual drainage systems by OTTSWMM. In *Urban Drainage Systems: Proceedings of the First International Seminar* (ed. R.E. Featherstone and A. James), pp. 2-93–2-108. London: Pitman Books.

15

Conference Conclusions

Richard N. Middleton

The closing roundtable discussion session of the Conference showed that there was strong consensus on two matters:

- low-cost sewer systems are an important technology which is cost-effective in many situations; and

- there is an urgent need to disseminate this message effectively.

However, the details of follow-up action remained, probably inevitably, somewhat contentious. Perhaps the debate can be simplified by dividing it into two parts:

- What are we trying to disseminate?

- How should this be done?

While this is an over-simplification (different lessons will need to be disseminated by different channels), it does at least provide a framework for decision.

The first step is therefore deciding what we know about the topic. There are clearly two conflicting approaches here. One is highly scientific and involves resolving professional differences of opinion over, for example, hydraulic design. The other is pragmatic, and says, in effect 'we know that sewers can be as small as 50 mm diameter and can be laid with grades as flat as 1 in 2000. Whether in theory these values could be slightly bigger or smaller does not matter. We cannot construct sewers more accurately than 1 in 500 anyway, and the saving in cost between 50 mm pipe, which we feel uneasy about, and 100 mm pipe, which we are comfortable with, is not significant. Adopting minimum values of, for example 1 in 500 and 100 mm as interim standards would enormously reduce excavation and pipe costs compared to conventional criteria'.

The biggest danger seems to be that acceptance is going to be seriously impeded if sceptics see the protagonists quarrelling over details which, while important, do not invalidate the general principle that revised design criteria are urgently needed and would result in great economies. The first 'message' to communicate to the profession in general therefore would be:

> Under a wide range of conditions, sewers designed to standards that reflect modern knowledge and which are carefully tailored to meet site conditions will be much cheaper, and just as effective, as sewers designed to conventional criteria.

In parallel, there needs to be a major effort to discover precisely what has been done in the past 35 years, and how well it has worked. This Conference has obviously been an important first step in this process. It needs to be followed by:

1. An inventory of alternative sewer systems, worldwide.

2. A compilation of alternative technical approaches and design criteria for the more important of these (importance being ranked primarily on innovation, not size).

3. An evaluation and comparison of the technical performance of systems representing alternative design approaches, distinguishing as far as possible between technical and institutional factors.

4. A synthesis of these findings (this needs to be an 'internal' meeting initially, so that any disagreements can be ironed out as far as possible before 'going public').

5. Presentation of this synthesis at an international workshop.

6. Endorsement of the approach by sector professionals and funding agencies, and agreement to initiate trials in those countries where it is unfamiliar.

7. Finally, implementation through a wide range of activities: country-level adoption of new policies; preparation of revised standards, model specifications, manufacturer's application guidelines, etc.; development of appropriate training and

educational materials; dissemination through professional meetings, etc.

A small but none the less important point relating to collection of past experience is the need to fund it fully. This means paying the people being interviewed as well as the interviewers! There is no doubt that in the past, while the academic community has published extensively (often as a condition of the research grant supporting the investigation), consultants have been much more reticent (because they are not paid for publishing, and tend to finish one assignment and move straight on to the next one), as have developing country specialists (because of language barriers or lack of access to editors of journals). Providing stipends for people who have vital operational information, so that they may either record it themselves or pass it on to interviewers, is therefore important.

It is interesting that almost no one at the Conference mentioned the need for evaluation of the institutional framework within which these projects have been undertaken. This obviously needs to be done as part of the fieldwork in steps 1 to 4 above, but it can probably be given less immediate emphasis during the next steps. It perhaps appeared that people felt that present fashion is maybe to focus too much on the 'soft' side of projects, while glossing over technical matters, leading to technical problems later (of course, the Conference was essentially a meeting of engineers ...). At any rate, agreement on the precise technical recommendations appears to be a precondition to any widespread use of the techniques; once the technical aspects are clearly understood, then the development of project-specific institutional arrangements can follow. These will almost certainly include some new types of partnerships, in which responsibility for design, implementation, management, and operation and maintenance is allocated between different types of institution [sector agency, private sector enterprise, non-governmental organizations (NGOs), the community affected]. Financial mechanisms will also have to be tailored to local circumstances (and will need to include household costs as well as municipal ones).

This research effort (steps 1 through 5 above) would probably take at least 2 years. The brief discussions at the Conference did not lead to any agreement on the ideal mechanism for carrying it out. Clearly, a number of different groups of people need to be involved:

1. The academic community, which has been involved in research into the basic hydraulics of the systems as well as in a number of pioneering pilot schemes, and which will, of course, educate the next generation of public health engineers.

2. Consulting engineers, who have designed and implemented low-cost sewer projects in a number of countries, and who have a significant influence not only on the standards and approaches adopted on specific projects, but also on broader matters such as national and international standards and codes of practice.

3. Aid agencies, which need to be much more insistent on appropriate technology in the projects they support, and which also need to ensure that their own staff are fully aware of the possibilities and limitations of various options.

4. Sector agencies, especially those concerned with sewerage, sanitation and urban development, but also including public health and environmental agencies.

There is a real danger of 'turf wars' preventing or delaying progress. In the international community, a number of agencies may feel that their mandate includes this topic (even though currently they may not be very active in it). These include:

1. The Collaborative Council Working Group on Sanitation (WHO, Geneva).

2. The VIP latrine group at WEDC, part of the Global Applied Research Network (GARNET).

3. UNCHS/Habitat, which has published an important guideline on the subject and which is the UN agency specifically charged with low-income human settlements.

4. The UNDP/World Bank Program, which has published several papers in succession to the original TAG Technical Note on settled sewerage.

5. UNICEF, which has now clearly stated its policy of inter-

vening in peri-urban areas, the most likely initial candidates for the adoption of low-cost sewer systems.

International units that could have important roles to play in the dissemination effort include IRC (WHO), SKAT, EHP (formerly WASH), ENSIC and the ITN managed by the UNDP/World Bank Program. There are of course many more at the national level, such as EPA's Small Flows Clearinghouse in the United States. The key criterion in dissemination has to be speedy and effective dissemination to the clientele—there seems to be a certain elitism in dissemination, which restricts access to publications and other materials (for example, by requiring payment in hard currency), and this should be resisted.

The choice of developing country partners is also critical. 'Hard hat' sewerage agencies may have little interest in adopting alternative design standards and practices, especially if this would require them to devote much more time to community relations. Lowering the cost of systems does not appeal to a number of influential groups—for example, agencies trying to enhance their budgets (and hence the size of contracts they can award); consulting firms paid on a percentage basis; or manufacturers of key components (especially pipes). Agency senior technical staff as well as senior university teachers may feel threatened by the introduction of design and implementation techniques which are unfamiliar to them, but which are well understood by their junior staff. The use of sympathetic, enthusiastic but non-technical agencies (such as community development agencies, NGOs, etc.) as the prime means of promoting applications in developing countries could also pose problems, as these agencies may not have the technical resources to ensure proper design, construction, operation and maintenance, and they may carry little credibility with national sector agencies. Ideally, the agency selected should be sufficiently influential to have the chance to change national policy, while at the same time being flexible enough to welcome new approaches.

15.1.1 Next steps

There is clearly a need to establish a permanent organization for promoting the adoption of more appropriate standards in sewerage design, and for assisting in the planning, design and implementation of low-cost sewerage. This valuable technology needs to be accepted as one of the tools available to solve the pressing problems of urban and peri-urban sanitation in developing countries. The most appropriate and immediate mechanism for such an organization appears to be to expand the existing GARNET system, to include a new unit concerned solely with low-cost sewerage. This new unit, however, needs to stress the 'applied' rather than the 'research' element of low-cost sewerage: enough is already known to apply the technology, even though future research and field studies will undoubtedly lead to refinements*.

The International Working Group should ideally include:

- members of the academic community actively engaged in research into low-cost sewerage,

- consulting engineers with field experience of the design and implementation of such systems in developing countries, and

- municipal engineers from developing country cities where such systems are in operation

Other people could be co-opted as necessary, to provide specific inputs. The group needs to be fairly small—perhaps no more than a dozen people. It needs to be financed, possibly by a bilateral aid agency, so that it has a budget for fees, secretarial support, publications, travel expenses, etc.

The prime responsibility of the International Working Group would be the promotion of low-cost sewerage. It would do this by preparing and distibuting a range of publications, and by arranging or conducting training courses in the various regions of the developing world. The publications should include design manuals (covering not only the hydraulic design of the sewers themselves, but also related topics such as the design of manholes, inspection chambers, and other appurtenances) and detailed case studies (including not only technical aspects but

*This new unit has now been established within GARNET as the Low-cost Sewerage Network; its co-ordinator is Professor Duncan Mara of the University of Leeds, who can be contacted at the e-mail address: d.d.mara@leeds.ac.uk.

also economic aspects, financial performance, sustainability, and institutional arrangements, with a particular focus on the role of the community). Techniques for widespread dissemination (such as distribution of videotapes or CD-ROMs, or setting up access through the World Wide Web) should also be explored.

Index